GW01086615

Good work, great technology: enabling strategic HR success through digital tools

Jo Faragher

Published by Clink Street Publishing 2022

First edition.

ISBN:
978-1-915229-49-6 - paperback
978-1-915229-50-2 - ebook

About the author

Jo Faragher is a business writer and editor, specialising in HR and employment issues. She regularly contributes news and features to *Personnel Today*, *People Management*, *First Voice* (Federation of Small Businesses) and *Times Higher Education*, as well as features published in national newspapers such as the *Financial Times*. Before becoming a freelance journalist 10 years ago, she worked as a content editor on *Personnel Today* and was magazine editor at the *Times Education Supplement*. Alongside journalism, she regularly supports corporate clients with copywriting on a host of business issues and covering events. She was awarded the Towers Watson HR Journalist of the Year prize in 2015 and was highly commended for the same award in 2019 and 2021.

Contents

PART THREE

EMPLOYEE EXPERIENCE AND ENGAGEMENT

PART FOUR

L&D AND CAREER PROGRESSION

Foreword

When I started my career, only the largest corporations used any form of people management technology (and they probably would have called it a 'personnel system'). Most employees would never access the software, which at the time were just simple databases for storing basic staff information.

Since then, the technology we all use – at work and at home – has gone through two major revolutions. The first was the proliferation of desktop PCs. When, in the 1980s, Bill Gates proclaimed Microsoft's vision to be "A PC on every desk and in every home," many people dismissed it as a fantasy. Now, of course, we can't possibly imagine how work ever got done without them. But their widespread adoption in the workplace still didn't mean that employees had much interaction with HR systems – instead, they would complete paper forms and HR would update their information in a central database on their behalf. Information was siloed and often inaccurate.

The second revolution was, and still is, the internet. From a people management perspective, the connectivity offered by the internet enables employees to take ownership of much of their own people data. Employees can access HR systems themselves and update their own details, understand their organisation's structure, manage absences, and view important documentation. This revolution was a result of, firstly, connectivity via PCs, and, later, via smartphones – meaning that even deskless employees can stay connected to their HR or people department.

Another, largely hidden, impact that the internet is making on people technology, is how applications are developed in the first place. With solutions being developed 'in the cloud', the cost of creating a new application has decreased dramatically. Unsurprisingly, in the last 10 years this has led to an explosion in the number of specialist software applications being developed to solve niche problems. At Ciphr, we have built thousands of integrations between our own HCM platform and specialist, third-party solutions. However, even now, a significant proportion of the integrations we build are to solutions that are so niche, it's the first time we've come across them. My favourite example is a link from Ciphr HR to a freezer management system that monitors how museum exhibits are stored. Who knew that was a problem that needed to be solved?

But that's really the point. Our view is that no single provider is ever going to be able to keep up with the pace of change and innovation across the entire people management technology market. Instead, it's better to focus on providing an excellent central solution, and connect it to the specialist software that is important to each organisation. In doing so, organisations can equip their people with the best tools to do their job, be as productive as possible, and deliver an amazing employee experience.

The challenge is that most organisations still prefer to procure all their people-management technology from a single provider – even if stakeholders in functions such as HR, finance and payroll aren't satisfied that a single platform will meet all their requirements. This is why we commissioned Jo, an experienced HR and business journalist, to write this book. We wanted to showcase some of the innovation and expertise that exists in the wider ecosystem of people technology. In this book, you'll find input from 93 experts from 87 different organisations, and this doesn't even begin to scratch the surface. A huge thanks to all of them for sharing their thoughts with us, and to Jo for crafting such a wide-ranging and inspiring book.

My hope is that, by reading this book, you'll view people-technology in a new light, moving away from attempting a one-size-fits-all approach, and instead embracing the possibilities offered by connecting the best technologies for your needs.

Chris Berry,
CEO and founder of Ciphr Group
June 2022

Introduction

HR technology has come a long way since the first purpose-built people systems emerged in the 1980s. But the rate of change has also accelerated since the days of HR software being little more than a digital record of your employees and their contact information. The dawn of the internet in the 1990s scaled things up rapidly, whether that was through the growth of job sites, the rise in employee self-service, or bigger and better intranets. And just as we were acclimatising to this, HR and employee-facing tools shrunk into the palms of our hands via smartphones and tablets.

These waves of evolution in HR systems have mirrored the changes to the function itself: from its roots in 'personnel' and administration – when the focus was on making transactions more efficient – to its more innovative and strategic role today where employee experience is king and HR's goals are closely aligned to those of the business. Speaking at the WowHR conference in 2021, HR professor and thought leader Dave Ulrich told delegates that "HR is not about HR; it is about success in the market and that value is defined by the receiver, not the giver."

But HR is now faced with an overwhelming choice of tools to help it deliver that value. The global HR technology market was worth almost US$23 billion in 2020, and is predicted to be worth US$35.7 billion by 2028,[1] according to analyst company Fortune Business Insights. The advent of cloud-based systems has freed up organisations from expensive on-

1 Human Resource (HR) Technology Market to Hit USD 35.68 Billion by 2028, Fortune Business Insights, June 2021, https://bit.ly/33lcdKe. Accessed 15/03/2022

premise implementations that take years to deliver a return on investment, and has opened up an ever-expanding range of add-ons to their central HR systems. This might be as simple as an applicant tracking system (ATS) to make recruitment processes more streamlined, but could be as sophisticated as a chatbot that monitors employees' wellbeing or a complicated algorithm that can predict when workers are considering moving jobs.

While the choice of specialist solutions available means it's possible for HR teams to pick and choose the best technologies for their organisation, this market fragmentation also creates some challenges. Research by one HR software vendor for example, found that 25% of HR professionals felt that too many disconnected tools were causing employees frustration,[2] while on average, organisations were using between six and eight tools to work on people-related tasks alone. Fads in HR technology come and go, so there's often a battle between flashy tools that make grand promises and knowing which ones will actually deliver improvements.

Tech 'spread' also places demands on HR teams' skills: people analytics requires data skills, for example (the profession's own industry body, the CIPD, has found that just 6% of HR professionals are using advanced analytical techniques),[3] while integrating multiple interfaces and tools together requires close working with IT teams to establish what's possible. Getting comfortable with these conversations is crucial to success when acquiring any new technology.

We've broken up the book into four areas covering what we feel to be the central pillars of HR technology:

- **HR administration:** the importance of a central HR system that can integrate seamlessly with a rapidly growing range of workplace tools. In this chapter we

2 Counting the Cost: How businesses risk a post-pandemic talent drain, https://bit.ly/3nqnPm2. Accessed 15/03/2022

3 People Profession Report 2020, CIPD, May 2020, https://bit.ly/3FCqD69. Accessed 15/3/2022

look at how the market for HR technology has evolved beyond a system of record to encompass remote working and collaboration technologies, sophisticated data analysis and automation of simple tasks. We'll also look at how payroll technology is evolving as the needs of the workforce change
- **Hiring:** it's the start of the employee lifecycle, and it's crucial to get it right. Recruitment technology has an important place in supporting HR teams when they're faced with a challenging labour market – whether that's finding the right candidates, building pools of talent for the future, or ensuring that candidates have the best possible experience of your brand before and when they join. How can technology match candidates to jobs? What is the role of video interviews? And how can we use the power of social media?
- **Employee engagement and experience:** how can organisations create a workplace where staff feel good, bring their best selves to work, and want to stay? As employee experience gains in importance, we look at how to measure it through sentiment analysis and surveys, as well as the role of wellbeing tools, social networking and consumerised interfaces
- **Learning and career progression:** workplace learning may have moved beyond 'chalk and talk', but how can organisations build it into the 'flow of work' in an age when it's easier to Google how to do something than take a course? We look at how organisations can assess the skills they need through continuous feedback mechanisms, and build this into their talent management strategies for the future. We also look at the evolution of content formats and the role of artificial intelligence (AI) in learning

Moving forward, it's all about how technology can enhance the experience of work for the employee. HR analyst Josh Bersin describes employee experience as a 'crusade' that every HR and IT department is focused on, which – put simply – refers to how organisations deliver a suite of tools for employees that helps them to not just be productive, but also protects their wellbeing and boosts their engagement with the organisation. The pandemic has cemented this shift from driving efficiency to making life easier and happier for the workforce: as companies pivoted to new ways of working, employees turned against managers who used tech to demand always-on availability, and the 'great resignation' proves many are now voting with their feet.

That's not to say that HR technology cannot drive business success by making things work more smoothly. It's an area ripe to be transformed through AI and automation, and organisations are already seeing success with AI in areas such as video interviewing and customised learning and development. Research in late 2021 from the Reward & Employee Benefits Association[4] found that 47% of organisations plan to change day-to-day workforce tasks through automation. At the time of writing, the HR tech market is growing at such a pace that many of the areas covered by this book will have evolved yet further by the time you read it. But as with every digital wave that came before, the key is to understand where it is you're trying to go, and how technology can help you get there.

Jo Faragher
June 2022

4 Technology change is business change, Reward & Employee Benefits Association with Mercer Marsh Benefits, November 2021, https://bit.ly/3qqXDde. Accessed 15/03/2022

PART ONE

HR ADMINISTRATION

How should we integrate our HR systems?

HR systems have evolved. While there will always be a requirement for a central hub of employee records, a shift to a more fragmented 'stack' of people-related applications has been going on for some time. As organisations look towards a future of work that is more about enhancing the employee experience and productivity of our workers; rather than just recording data, they're looking at a host of specialised solutions that can help them do that. And employees' expectations are high too: whether it's onboarding or a digital performance management process, they expect this to happen seamlessly.

"Organisations are looking for ways they can increase productivity, but also enhance the employee journey and create a new employee experience," says Matt Russell, chief commercial officer at Ciphr. "So if you need to onboard someone virtually, how do you get them to feel part of something? If people are not in the office, how are you helping them understand the culture, engage them, and help them become productive team members as quickly as possible?" To create that seamless experience, HR systems need to be able to connect with other systems used across the organisation that rely on accurate and up-to-date information about the workforce.

A survey by Ciphr in 2018 found that more than half (51%) of HR professionals felt that there was no single piece of HR software that would be able to satisfy all of the HR requirements for their business. In addition to their central HR system, 44% said they used a separate standalone payroll system, 23% a time and attendance or rostering system, and 21% a separate tool to manage expenses. Other systems in HR teams' arsenal include performance management software, background checking systems, employee engagement platforms and learning management systems (LMS). But how seamlessly those systems integrate is an important consideration. According to Sven Elbert, senior analyst in the HCM practice at Fosway, eight to ten integrations is typical for a medium-sized company, while very large organisations could have up to 700 connected to their central HR system.

Why is integration so important? Connecting systems together is not only about having 'one version of the truth' to provide accurate data insights – it also has significant benefits in terms of efficiency. Seamless integration can streamline workflows for new starters and leavers, and when employment terms are altered, because applications will be automatically updated with this information; there is less admin work for the HR team; the risk of data duplication or mistakes in data is lessened and it makes compliance with data protection rules such as GDPR simpler because it's easier to see who controls and processes the data. Beyond reducing admin, integrations can make people data available to other systems in real time, enabling managers to make better-informed decisions more quickly.

"It often starts off with an organisation looking for a system that does everything," says Helen Armstrong, managing director of integration specialist Silver Cloud. "Perhaps they want to reduce the number of contractors they're dealing with. But there's not one software vendor that does it all: for example, if you do a lot of recruitment, we might advise a standalone system that would work well with your central HR system."

Building the business case

Research suggests that organisations are spending more on HR technology than ever before. A survey by consulting firm PwC found that 74% of companies planned to increase their spending on HR technology in 2020, on top of the around $310 they already spent on it per employee per year.[5]

One of the reasons for this increased investment is that there is simply far more to choose from. Whereas a traditional HR stack might once have incorporated employee records, sickness absence and an ATS for recruitment, organisations can now invest in anything from wellness trackers to employee sentiment mapping tools.

Furthermore, cloud computing has made investing in software both cheaper and easier because organisations no longer need to integrate with on-site hardware, but that doesn't mean it's easy or cheap in every case.

Identifying how the 'ecosystem' of HR and related software should look, and the type of integration between different systems will depend on a number of factors:

- The nature of your business
- How sensitive the data is to loss or exposure
- How time sensitive the data is – do you need it in real time or just once a week?
- How seamless the experience needs to be for the user
- Budget
- IT resource and the capacity of your internal tech support team
- How the integration will be used. Is it just updating data, or will it be used to trigger workflows?
- Which tasks are a priority
- Future plans for expansion, including acquisitions and opening overseas locations

With so many considerations, opting to invest in the best application for its purpose and integrating it with your central HR system might seem counterintuitive, but there are good

5 PwC's HR Technology Survey 2020, PwC, Sep 2019, https://pwc.to/3tllzjX. Accessed 15/03/2022

business reasons. Elbert from Fosway adds: "If you keep these systems separate you have more people doing admin work or keying in data more than once, which takes time and resource and leaves room for potential errors. The main differentiator between systems we see now is not functionality – the bells and whistles – but their capability to link to other systems."

The market for employee-focused software is growing, but how well these tools connect to central HR systems varies widely. "With some connectors you can just click a button and have a connection, but less sophisticated ones may require more technical support. This is why many organisations approach companies that specialise in integration services, or invest in 'middleware' that can sit in between the two systems and connect them," he explains.

The ability to 'bolt on' applications carries many positives, however. It means organisations can add tools and users more flexibly as their business grows (or contracts), and enables them to bring in specialist tools for ad-hoc projects. With employers relying on increasing numbers of contingent and freelance workers, the flexibility to bring them into the business and access tools securely will also become more important.

Managing vendor relationships

Using multiple systems means managing multiple relationships and contracts with the vendors that make the software, and this can deter some HR teams from integrating best-of-breed systems. Russell notes: "Customers know that best of breed can have challenges; there are two sets of invoices and possibly two different subscription periods, or perhaps they've had a bad experience of having to use a particular system before. But they will accept the complexities associated with managing multiple suppliers if the systems solve a critical problem (or problems), or achieve the end result they want."

Denis Barnard, director of HRmeansbusiness, a consultancy, says it's important to define how investment in multiple tools will align with the wider goals of the business. "You don't start off with 'let's buy an HR system': organisations should build a shopping list of features that they need. This might be payroll, time and attendance, and self-service apps for employees," he says.

If you decide to integrate a mix of best-of-breed software from multiple providers, it's important to interrogate suppliers on their integration claims. Some will say their connectors are 'plug and play' but actually require extra work behind the scenes to make the systems interact as you want, or perhaps you need a level of customisation, but the integration will only work in a certain way. Armstrong adds: "Integration can mean many things; it could mean you get an upload of data once a month or once a week, and so this should be part of the conversation. And when vendors say they can integrate [their software], they might just mean doing so by importing an Excel file." Many vendors will have a list of trusted partners that integrate more readily with their system: Ciphr Connect, for example, has a partner marketplace where customers can see applications that have two-way integrations to Ciphr's central HR, recruitment, payroll and learning systems.

HR professionals do not require deep technical knowledge to manage these relationships as the technical aspect will best be managed by the IT team. The key thing is that HR should be able to map out why and where different systems should talk to each other. What is the benefit in exchanging data between these systems? Does it have to be every single piece of data or just one aspect? How often should this happen? Russell adds that external implementation partners can help make those connections a reality, as well as discuss potential scenarios with customers and how these might help users do their jobs more efficiently.

User experience and expectations

There is more of a focus than ever on user experience when it comes to workplace technology. Employees expect that the technology they use at work should be far more advanced than what they use at home, but this is not always the case. At the very least, they expect a system to be easy to sign in to, or not to have to switch between multiple applications, so they can get on with their core work tasks.

Their experiences as consumers inform this, says Armstrong from Silver Cloud. "Many workers are from a generation that's used to doing everything from banking to booking a holiday online and expect the same sort of experience at work," she says. "The company's drivers for investing in integration may be to reduce paper processes or make things quicker, but you also need to think about the user experience."

These expectations have only accelerated during the pandemic, when employees became more used to performing tasks online that they used to do in the office. "From an employee experience perspective, the organisations that thrive will be those that are really clear about what it means to work there, so it's not just about streamlining processes but what the experience feels like," says Russell from Ciphr.

Looking forward

A 2020 survey conducted by the World Economic Forum found that 80% of organisations plan to increase digitisation of operations and expand their use of remote work, with 50% intending to accelerate the automation of certain tasks.[6] The UK's Office for National Statistics, meanwhile, found that employees that were able to pivot to remote working during the pandemic actually drove up productivity levels last year, suggesting there are productivity gains to be had from effective digitisation of processes.[7]

---

6 The Future of Jobs Report 2020, World Economic Forum, October 2020, https://bit.ly/3rz6OaT. Accessed 15/03/2022

7 Productivity economic commentary, UK: January to March 2021, Office for National Statistics, July 2021, https://bit.ly/3qnsirx. Accessed 15/03/2022

At the same time, however, we have seen a shift towards more employee-led decision making and an increased focus on smoothing the user experience – whether that's while working remotely or in the workplace. As this continues, the market for potential 'add-ons' to central HR systems will grow and grow. If your HR team wants to get the most out of its technology ecosystem, invest wisely and ask the right questions to reap the benefits.

How does integration work?

Many organisations have multiple systems that feed into HR operations and decision making, often meaning the same data is kept in multiple locations. A typical mix might be a central HR system linked to a payroll system; there might be an ATS feeding in recruitment data; and potentially 'peripheral' tools such as employee feedback. But how does this 'ecosystem' link together?

Sven Elbert, senior analyst in the HCM and talent management practice at analyst group Fosway, believes there are four types of integration:

Importing – where adjacent systems send batches of data via Excel, CSV or text files, perhaps on a nightly basis.

API (application programming interface) – where systems from different vendors offer interfaces that can 'talk' to other systems. In short, this is a series of rules and protocols that help applications fetch data from other systems. Many vendors will create APIs that are unique to your organisation, or, if you have in-house technical expertise, it's possible to customise how systems communicate.

Standard connectors – where vendors have built adaptors between different systems that mean organisations can, in theory, 'plug and play' a connection between them.

White-label partnerships – where vendors have coded in data flows already so the interaction between systems is invisible to the user.

The type of integration that works best will depend on the reason for connecting each system. Some activities will require real-time connectivity between the tools, while for others a nightly dispatch of data will be perfectly adequate. There may also be considerations around data security and GDPR, where minimising the number of systems or people that handle the data is crucial.

Five key takeaways

- Understand the business benefits of integrating other systems with your central HR software, such as data security, less duplication, improved efficiency and a 'single view of the truth', all of which ultimately provides you with more robust and accurate data and business intelligence
- You don't need to be a tech expert – but you need a clear idea of what HR wants and needs to achieve from linking systems to each other
- Decide why systems need to talk to each other and the level of data they need to exchange, and how often
- Challenge vendors on claims of 'plug and play' integration – is this a line of code or will it require extensive customisation from your IT department?
- User experience should be the priority – employees increasingly expect a seamless transfer between systems to help them be productive

Jo Faragher

Analytics, modelling and forecasting: using people data for insight

When the pandemic turned business as usual upside down in early 2020, HR teams were faced with more questions than they had answers for. Did their organisation have the capacity to support the shift to working from home? How would furlough impact pay and staffing levels? How could essential departments deal with potentially high levels of sickness absence?

Research carried out by the HR Analytics ThinkTank, a partnership between universities and private companies that compiles intelligence on the people analytics field, shows that data played a crucial role in decision making during the Covid-19 crisis, helping HR teams address immediate issues and answer unprecedented questions. Nigel Dias, the think tank's founder and managing director of consultancy 3n Strategy, explains: "The pandemic saw businesses making decisions they've never made before, and there were questions no one had ever answered before. People were either guessing, or they had data which enabled them to make better choices."

People analytics – the discipline of using HR and related data to make organisational decisions and even predict outcomes – will continue to play a key role in decision-making as organisations recover from the pandemic, the think tank concluded. "The next stage will see people analytics teams answer new questions, including questions about how employees are coping with their new working arrangements, including the challenges of working from home, risking their lives as essential employees, or leaving their employment – and the overall mental health of employees as well as dealing with the uncertainty and stress of the situation."

For many HR departments, however, the pandemic represented a shock to the system in terms of the data they had available and their capability to deal with it. Consulting company McKinsey identified four key uses for people analytics during the crisis: workforce sentiment analytics, looking at how people felt; workforce protection data, such as social distancing or how many employees are in the office; visibility of who is working on what remotely; and workforce availability. To make robust decisions in these areas, HR teams often needed to access reliable and real-time data feeds from multiple systems outside their central HR system, such as recruitment and rostering systems, and employee engagement or surveying tools.

The pandemic also brought into sharper focus the debate around HR professionals' skills in people analytics. "In the past 10 years the companies that do this really well, such as Facebook and Google, haven't really changed," says Professor Andy Charlwood, chair in human resources management at the University of Leeds. "There's a feeling among HR that they're good at managing people so they question the value of data. Either that or they believe their HR system is there to make processes more efficient rather than create insights." There are also mundane practical issues such as staff setting up fields in different formats so data cannot be compared, or failing to input data at all. "Sorting out the data is a big job and not a nice job, and so it seems to become more arduous than it actually is," he adds.

A survey by the CIPD carried out in 2018[8] found that only just over half (54%) of organisations had access to people data and analytics, and only four in ten felt their HR team would be

8 People analytics: driving business performance with people data, CIPD, June 2018, https://bit.ly/3fjq8mH. Accessed 15/03/2022

able to tackle business issues using data analysis. Three-quarters who had access to people data were using it to tackle workforce performance and productivity challenges, and in organisations where respondents felt their analytics culture was strong, 65% reported better business performance compared with competitors. The CIPD advised a cross-functional approach to improving HR's skills and confidence in this area. "There are clear differences in the perspectives of HR and finance professionals, and other professionals using people data," said its report. "Non-HR functions need encouragement to increase their use of people data in their decision-making and HR has a role to play in generating trusted, relevant people data to serve wider business needs."

Sources of people data

Organisations can draw on a range of data sources to model how their workforce looks against different parameters now and in the future. These include but are by no means limited to:

- Central HR system data (basic information about individuals, such as name, job title, department)
- Demographic data (from the central HR system and through disclosure of, for example, sexual orientation, socioeconomic background and ethnicity)
- Recruitment systems (for example, how many hires, time to hire, and diversity of new hires' backgrounds)
- Onboarding data (eg drop outs, time to productivity, targets for training)
- Learning management systems (eg course completion, skills needs, popularity of courses, compliance)
- Employee surveys (eg in which departments are people most satisfied? Who needs support?)
- Absence records (if not held in the central HR system)

- Performance management tools (such as feedback scores or comments)
- External sources (such as job boards, pay benchmarking, and social media)

Professor Charlwood recommends organisations that are starting their analytics journey begin with the business problem, rather than expecting the data to present solutions. "Take sickness absence, for example. It's easy to see who is off sick, but not necessarily know why. Perhaps the data in the HR system shows that it is women in their late-30s who could be taking time off to look after children, but is that actionable insight?" he says. "To get to the insight you need to collect data in the right way, so for absence that might be through return-to-work interviews. Once you know the answer, that can inform how you change how employees are managed or their work is designed."

Spending time at the start of any data analysis project identifying the right questions, then defining the types of insights the organisation would like to see from the analysis is crucial. Dias adds: "Anything HR achieves is a result of a decision it has made, and at any time HR professionals could be making hundreds of people-related decisions. If your gender pay gap is X, then what decisions were made to get there? If you want to reduce it, what is the factor that will make the change?" One of the barriers can be that seasoned HR professionals feel that they have been making these decisions for years and "don't want to feel that a piece of analytics is more powerful than their expertise." The key, says Dias, is in understanding that the data can help teams make better choices rather than replacing their years of experience.

One area where the role of analytics has come to the fore during the pandemic is employee experience. Suddenly, organisations craved more insight than ever before into how employees were feeling, their stress levels, and their physical wellbeing. Many opted for more regular pulse surveys on topics such as mental health, producing a wealth of data points for analysis. Rob Robson, director of people science at the People Experience Hub (PX Hub), says the same principle applies as with any strand of data. "Why do you want to collect data from employees in the first place? What problem are you trying to solve? What outcomes do you want? Going straight into investing in a platform could close off options. Your research question should be your mantra; you need to determine what you challenge is, before you think about how you're going to solve it."

Collecting data from employees about their experience, or asking them to disclose additional personal information, can feed into useful insights on improving the business culture and diversity and inclusion – but it also requires careful handling. Data protection regulations (the GDPR) require organisations to be transparent about how and why they are collecting personal data, and to minimise the number of people who need to handle it. Informing employees how their data will be used, presenting the insights to them, and showing that they have been acted upon can help build trust. Aggregating and anonymising any datasets from particular groups ensures individuals cannot be identified, adds Dias, but often the organisation's culture will drive how receptive it is to using people data to drive decisions. He adds: "Engagement data can be hard to marry up with other strands because it's anonymised, but the biggest factor in building a data-led culture is leadership. If you don't have sponsorship at the highest level, every other measure will be undermined."

When it comes to modelling people data, how systems are integrated beneath the surface will influence the efficiency and reliability of insights for HR. There are a number of tools on the market, such as Visier and Capterra, that will draw data out of different systems and automate the processing. Microsoft's Power BI tool, for instance, will generate data visualisations and insights from CSV files. "If you've integrated systems so that data is up to date in real time then you can make evidence-based decisions based on the data you have. You're running a board report that's up to date, rather than working from an old spreadsheet," explains Russell from Ciphr.

The key to unearthing deeper insights can often lie in overlaying data from different sources. At a simple level, this could be marrying up data on attrition in a certain demographic group with the same group's engagement scores. Building it up step by step can build confidence, adds Russell. "Start with a question such as 'How do candidates rate the recruitment process?' and once you've got that, you could look within that at age, gender, hiring manager, and interview stage. HR systems collect a huge amount of data and it's a case of getting it all in one place."

Robson adds: "It's about looking for patterns in the data, and working across different indices to see where something might have an impact on, for example, your change strategy." A culture that is supportive of data-led decisions is crucial, however. "If you identify a problem in the recruitment process using that data, for example, will it be taken on board and addressed?" asks Russell. "What will be the action based on what you have found?"

The next step is using data from your systems to predict future outcomes – a practice that is already established in areas such as marketing and engineering, but can be more problematic in HR. "Statistically, it's more complicated," says Professor Charlwood. "A marketing team, for example, might want to know which version of a website makes people buy more stuff, or in a factory you might want to predict when a machine will need maintenance. In HR, predicting whether a candidate will perform well and making a decision based on that could expose you to legal risk. Or the data could suggest that 20 people are going to leave, but which 20? It's much more complicated with people than machines."

Instead, he advises using the suggested insights from the data to augment future decisions rather than make them for you; improving access to wellbeing support if a data model predicts attrition due to burnout will benefit everyone, for example. Often it's a case of spotting a potential anomaly that provokes further investigation. "The data might not give you an answer but will give you the question to ask, a red flag or conversation starter," he adds.

In a time where making reliable decisions is a challenge and the workplace context is constantly shifting, HR teams will need to rely on data more than ever before. In the months and years to come, the post-Covid workplace will throw up questions about safety, productivity and wellbeing, many of which can be answered by data in HR systems. But it's also an opportunity for HR professionals to improve the decisions they may previously have made based on intuition or experience. As Robson concludes: "Interpretation comes from the tools, but at a more strategic level you can't replace human judgement and the application of experience."

Five key takeaways

- The effect of the global pandemic has accelerated the need for data-led decision making and this will continue to evolve
- Effective integration of HR and associated systems can improve data quality and the level of insight produced
- A data-driven culture can boost HR's confidence to use evidence-based insights, particularly when these insights are acted upon
- HR professionals should talk to colleagues and peers in other functions about how they use data, or the benefits they could get from using people data
- Predictive analytics can suggest potential scenarios, but should inform decisions rather than make them

How can technology help remote collaboration?

When HLM Architects went into lockdown at the start of the pandemic in 2020, much of its communication transferred to Microsoft Teams. "It was a familiar interface so not completely new, and at first we'd use it for client engagement and to discuss designs," says Marcus Earnshaw, technology director at the firm. "It was a good experience but not quite the same as being together. Usually we'd be sitting in a room around a blank sheet or doing mark-ups and taking notes." In response, Earnshaw and his team began looking at ways they could use Teams more collaboratively, adding whiteboard functions so designers can make comments on a 'virtual' sheet of paper, making more of the chat function, and combining this with video meetings.

As the company moves towards a hybrid working model, this new approach has heightened rather than reduced its collaborative capabilities, Earnshaw believes. People from other offices don't have to travel for meetings, amendments can happen quickly, and employees who feel less comfortable speaking in meetings can switch to Teams chat instead, so it's more inclusive. HLM wasn't alone in expanding its use of remote collaboration technology over the course of 2020 and beyond: according to annual rankings by software company Okta, three of the five fastest growing business apps in 2020 were Miro, Figma, and Monday.com, which all fall into this category.[9]

9 These are the fastest-growing corporate apps, Fortune, January 2021, https://bit.ly/3394Ci8. Accessed 15/03/2022

Carl Harris, group director at BCS, the Chartered Institute for IT in the UK, says: "Certainly, at the start of the pandemic, the single biggest change in our use of collaboration tools was not which tools we are using, but how we would use the ones at our disposal differently." One of the best examples of this is Microsoft Teams, which went from being an additional communications channel to an "integral part of our everyday working… In the past we may have considered that these types of tools had more features than we could usefully take advantage of – now new features and improvements from the tool's development roadmap can't come soon enough."

Most employees will now be familiar with tools such as Teams and Zoom for rudimentary workplace communication, such as setting tasks or asking questions. But there is a host of other collaboration tools available, such as:

- Virtual whiteboards: users can place sticky notes, add pictures, draw diagrams, and include links on a central 'canvas'
- Project planning and management tools such as Trello, Asana, and Monday.com
- Interactive spreadsheets: cells have data but can also link to other resources, and cross-reference with other sheets or documents. Sheets can be edited in real time so there is always a single version of the truth
- Video collaboration: enhance Teams and Zoom functionality by adding the ability to edit a document during the discussion, or enabling screen sharing
- Dashboards: a visual means of showing who is working on what or how many people have responded to a task, for example

"Learning how to use the tools is one challenge, and knowing how to use tools efficiently and productively is another – but the new third dynamic is knowing how to use these tools optimally in a hybrid working environment," adds Harris. He points out that one of the difficulties for organisations moving to a combination of remote and office-based working has been that – while these tools often make people more productive – it's harder to maintain that feeling of togetherness, or culture. "As effective as tools have been in facilitating remote contact, they fall short when it comes to facilitating spontaneous communication between colleagues; the 'water cooler' conversations are so hard to replicate in a virtual context. There are also the new challenges of managing feelings and symptoms of isolation among employees who are working virtually. All these factors put pressure on tools that may not have been designed with the hybrid working culture in mind."

Building the culture for effective remote collaboration is arguably what will define the success of the tools behind it. Starting off by using collaboration tools in a social or nonformal way can help – it can get employees comfortable with using the interface before introducing additional functionality, while enabling employees to recreate some of those serendipitous conversations virtually. "Using check-ins or warm-ups is a way of getting employees used to tools and into the mindset of digital collaboration," says Jim Kalbach, chief evangelist at virtual whiteboard company Mural. He argues that the visual component of the new generation of collaboration tools changes the game in terms of remote participation. "People can express themselves visually, draw on something, and put an arrow to show a relationship between two things," he adds. There is an added bonus in terms of inclusivity, too. A recent paper in Scientific American[10] argued that 'virtual brainstorming' can actually be more innovative and provide a better experience for all group members than in-person meetings, because more introverted employees' contributions aren't blocked by others' noise and there is less focus on the seniority of those involved.

10 Remote Work Can Be Better for Innovation Than In-Person Meetings, Scientific American, October 2021, https://bit.ly/3K8MhC7. Accessed 15/03/2022

Halbach agrees: "It gives a voice to people who might not otherwise participate, changing the dynamics of the meeting and creating a different power structure – one where an intern has as much space to talk as a CEO."

One of the risks of introducing new modes of collaboration into the workplace can be overwhelming staff – too much digital 'noise' in the form of endless Slack chats or WhatsApp threads can have the opposite effect of making them less productive. Ross McCaw, CEO and co-founder of OurPeople, an internal communications tool, says the trick is to get the right information to the right people at the right time. "People get fatigued with endless communications, when 90% of it has nothing to do with them. You don't want to have to scroll through lots of messages to see the one thing that's relevant to you," he says. OurPeople was initially designed for non-desk-based staff, so sends short 'cards' containing information (which could be a task, a questionnaire, or a shift request, for example) only to those employees who need to see them. "It's limited to 700 characters because that's the best view you'll get on a smartphone," he adds. "If you send information in digestible chunks, rather than *War and Peace*, engagement is much higher."

With multiple people working on a project from different locations a 'single version of the truth' is essential, so transparency and real-time updates are crucial. Ralf Dyllick, founder of interactive spreadsheet company SeaTable, says managers can build in permissions for certain employees across many collaboration tools so adding to a piece of work doesn't become a free-for-all. "You could be planning a marketing campaign, presenting the plan just as though you were standing in front of a whiteboard. All of those changes are logged and you can revert to the original if something goes wrong," he explains. An extra bonus is the data produced by collaboration tools – knowing who has contributed to what and when means there is greater transparency around what individual employees are doing, he adds.

Long before the pandemic, tech companies had been using virtual project tools such as Monday.com and Trello as part of a shift in culture towards agile working. The idea is that teams work in smaller 'sprints' and meet smaller goals more quickly, with the understanding that things can always be changed in the next iteration. The move towards hybrid working practices will see agile working become more widespread, argues Kalbach. "These tools feel very natural in an agile working environment but we're increasingly talking to HR teams and sales teams about how they can use visual collaboration to solve their own problems.

"The most challenging aspect of remote collaboration is in how we use tools to augment human interaction when employees do come together," he adds. "Whether we liked the physical workplace or not, there was intentionality about how we collaborated. We need to grab onto that and use this as an opportunity to transform how we collaborate at a deeper level." If someone has a contribution to an ongoing project, if they think 'digital first' about how they will add their suggestion this becomes part of day-to-day habits and therefore culture, he suggests. Dyllick at SeaTable believes we're some way off being able to replicate a feeling of togetherness with software, however. "The holy grail on this is still to be found – in the hybrid world we still need to find ways to instil that 'we' feeling, and it's not going to come from social media 'likes'. Identification, for example if you're a new starter, can only come from direct interaction."

HLM Architects recognises this, so the aim is to "always put the person at the front of the system, rather than a tool that would remove the human element," according to Earnshaw. So Teams is not only used for co-design but also for employees to check in to work (remotely or otherwise), and for wellbeing initiatives such as step challenges and blocking out time periods for a 'virtual commute'. "It's about getting the blend right – so if you want to convey information there's one tool, if you're working on a project together there's another," he says. "We need to make sure we don't have lots of tools that essentially do the same thing, so you're getting pinged from every direction. We want to remove the noise, and streamline what we do." Like many organisations, getting the mix right for remote and hybrid collaboration will take some trial and error, but this is one area of workplace tech that will grow and grow.

Five key takeaways

- Reduce the noise: overloading employees with multiple apps and feeds could risk them feeling overwhelmed
- Collaborate more inclusively: remote interaction gives more equity to voices that may not be heard in traditional meeting situations
- Go visual: people process visual stimuli and connections more quickly. Use a combination of text, pictures, links, and video to plan projects
- Ensure transparency: most tools allow you to set permissions so the right people can update a piece of work in real time and there are no other versions or loss of data
- Create an audit trail: the benefit of collaborating remotely using digital tools is that there is a digital history of who worked on what, which can be used to inform career discussions or to plan future projects

How can HR embrace the benefits of automation?

The early 2020s have produced a perfect storm for employers, from dealing with a pandemic to mounting skills shortages, increased wage bills and hybrid working. It's perhaps no surprise that, according to a survey by consulting firm EY, 41% of businesses plan to accelerate automation as a way to replace jobs in the coming years.[11]

In the HR function, self-service has been on the increase since the late 1990s when business professor Dave Ulrich popularised a 'three-legged' model of HR which pushes much of the administrative side to shared service centres and passes more responsibility for basic tasks (such as approving time off) to managers and employees. But increasingly, robotic process automation (RPA) and AI technology could replace a much greater range of repetitive tasks: a survey by Willis Towers Watson[12] with the Singapore Ministry of Manpower found that 24 out of 27 HR roles would be impacted in some way by automation. This could mean the end of some entry-level administrative roles, it argued, and for more senior HR professionals a way to escape repetitive administrative tasks and focus on higher value activities.

RPA tends to work best with repetitive, rules-based tasks. A computer 'bot' can learn the steps in the process and be given conditions under which it does one thing or another. This could be

---

11 Global Capital Confidence Barometer, EY, March 2020, https://go.ey.com/3rj7aSL. Accessed 15/03/2022

12 ow will technology impact HR jobs and skills?, Willis Towers Watson, December 2020, https://bit.ly/3nkFITy. Accessed 15/03/2022

via a consumer-friendly user interface such as a chatbot, or an engine that works away in the background. Fortunately, you don't have to be a technology expert to put RPA to work in your HR function. There are companies that specialise in RPA such as UiPath and Automation Anywhere, that can help with process mapping and offer a menu of prebuilt automations organisations can buy off the shelf. For those used to Microsoft platforms, its Power BI tool can automate simple data-related tasks such as populating spreadsheets or distributing reports. You might also find that your HR system already enables you to automate some key HR tasks.

Some examples of how automation might be used in the employee lifecycle include:

- Recruitment: booking interview times, screening applications/CVs, scanning job sites, posting ads, and sending out assessments
- Onboarding: arranging computer equipment, linking start date to payroll, setting up email accounts, alerting managers about new starters, and sending out communications about company
- Pay and benefits: managing paid leave, approving holidays and expense claims, timesheets and scheduling, and payroll calculations
- Performance and engagement: sending reminders for reviews, sending out pulse surveys and compiling results, and sending alerts related to HR cases such as grievances
- Offboarding: scheduling exit interviews, updating records, issuing the final payslip, removing access to applications, and removing an individual from the employee directory

Recruitment is a natural fit for RPA because so many aspects of the process can be time consuming and rules driven. A study by automation software company Workato[13] found that the use

13 Study finds pandemic led to sharp increase in HR automation, Unleash, September 2021, https://bit.ly/3zTsqlQ. Accessed 16/03/2022

of recruitment automation grew 547% over the course of the pandemic, compared to 257% growth in use for automation in HR services and 242% growth in learning and development automation. Onboarding and offboarding staff makes up for 34.5% of all automated processes, according to Workato. "Lots of automation happens in recruitment without being obvious," explains Anthony Wheeler, author of *HR without people? Evolution in the Age of Automation, AI and Machine Learning*. "It's not just scraping candidate data from the web – it might be gamified web tests or something testing a candidate's decision-making without them realising." Early-adopter organisations are moving to automate first-round interviews by using AI bots to ask questions or asking candidates to record a video response based on a series of questions, which will then be reviewed by an algorithm.

Since HR is the 'people' function of the business, it may seem anathema to many organisations to automate any aspects of the employee lifecycle. But, because "HR departments often make the 'human' part of their department a point of pride, this can translate into a mentality where HR professionals are used to rolling up their sleeves and getting things done themselves, which could make automated processes seem like cheap ways to get around doing the hard work," says Mel Kasulis, project manager at Skynova, which specialises in technology for small businesses.

However, adds Kasulis, "HR is also typically one of the best-suited departments for automation, given the high volume of mundane tasks HR professionals have to tend to." Some of HR's reluctance to embrace automation is driven by fears that the technology is too complex, when in actual fact lots of modern RPA solutions don't require many new skills, she says. "They're made to be easy to use without much technical knowledge once they've been properly configured. Once the automation is in place and paperwork is taken care of by the product, HR departments can focus on the human part of their job: supporting employees directly, hearing their concerns, and helping them find solutions to their problems."

Before embarking on any automation journey, HR can build strong foundations for success by working with other departments and involving employees in the process itself. "Automation works best where there is a core team of interested and invested people," says Len Pannett, a business transformation consultant. "Where it fails is where people assume processes are fine but they actually have a lot of variation, or where they don't think about how they will scale up from an initial pilot." Culturally, getting employees used to having aspects of their jobs automated can be a challenge, so transparency helps. Pannett adds: "The last thing you want is your RPA team doing this in a room somewhere where no one can see them. Sit them in the middle of the floor, and once employees see boring aspects of their role taken away, they see how good it is and start making suggestions." From here, it then becomes a case of prioritisation and process management, he advises.

Once automated processes are up and running, the return on investment can be quick and significant, adds Pannett. This is where teams will need to make decisions about whether they invest their budget in new hires to take over basic functions such as data entry or CV scanning, or scale up their technology. "With RPA, you don't necessarily need to hire another person if the size of your workforce doubles. A bot doesn't differentiate between 10 repetitions and 2000," he says. It makes sense to train up employees in different areas of RPA (such as process design, process management and system integration) because it can be highly scalable. He adds: "When this is done right, return on investment can be measured in weeks and months. We're seeing marketplaces of preconfigured bots emerge that can conduct basic processes such as pulling data from a database. Or larger companies are creating one RPA and using it many times across the business."

How does this impact the human side of the HR role? Professor Adrian David Cheok, professor at the i-University in Tokyo, argues that AI can augment people professionals' decision-making on top of automating the more mundane aspects of the role. "The decisions around someone asking for leave or calling in sick are not so complicated because if you feed the AI the rules, it makes the decision for you," he says. "But it can also pick things up early, such as patterns where people are taking sick leave frequently, prompting HR to have a conversation about the absence." As automation and AI becomes more sophisticated, organisations can programme more flexibility into them, adds Professor Cheok. "You could create a +20% wiggle room for something or program in more conditions that makes it 'friendlier', or you can make it more rigid, depending on what you're using it for," he says.

According to analyst company Gartner, the next step in RPA will be 'hyper automation': a combination of tools that augments an employee's job rather than simply replacing aspects of it. Hyper automation would see employees becoming 'citizen developers', it argues, delegating tasks to their robot 'twin' and using other aspects of AI and machine learning to improve other parts of their role. RPA company UiPath argues that anything in an organisation that can be automated should be, and, by training staff to think and act as developers, employees can readily see any process flaws and fix them. In response to this, we'll see organisations build centres of excellence to oversee the automation of processes across the business, says Pannett. "To make this sustainable, organisations will need to build up a way of managing RPA – where someone looks after not just the technical pieces of the transformation, but also the business case, how you manage updates, how you test processes and the like."

One of the major challenges in embedding automation will be the ethical considerations around what can – or should be – automated. Wheeler believes that, despite high investment in automation and AI, many organisations are still grappling with the ethical issues around the impact automation might have – not just on the workforce, but on decision-making. "Companies are asking questions such as: the technology is here but should we use it? What does this mean for our workforce?

What is the safety net to deal with technology displacement?" Multinational companies may also find the shift to automation difficult as different countries may have varying legacy systems or different local attitudes towards aspects of certain roles being replaced. "For HR, this means their role becomes highly strategic, because in some areas they may be operating at high speed, while in others, they face a more traditional infrastructure," Wheeler adds.

He concludes that the HR professional of the future will need to become adept at being the interface between people and technology – identifying where an organisation deploys technology, where they deploy humans, and where humans and technology can work together. "[HR will] not only be a strategic business partner but a tech business partner, working in really complex environments," he says. It's a daunting opportunity, but one HR professionals will need to embrace.

Five key takeaways

- Start small and simple: repetitive and rigidly defined processes work best
- Identify ways to scale up: once a few processes have been automated, how can you apply this to other areas of the business?
- Take employees on the journey: be transparent about what you're automating and engage employees in the suggestion and development proces
- Upskill employees: seek out training opportunities in aspects of RPA such as business process management and integration
- Consider the ethical implications: just because something can be automated, doesn't mean it should be

Why getting pay and reward right is more important than ever

Pay is the one area of the people function where there are rarely prizes for getting it right, but you'll soon hear about it if things go wrong. "There are no grey areas like there are in HR. It's right or it's wrong and employees expect 100% accuracy. There are not many roles where that's an expectation," says Amanda Barnden, payroll sales manager at Ciphr. Furthermore, it might not just be a case of an administrative error meaning an employee earns a little less one month – noncompliance with wage legislation can lead to investigations by HM Revenue & Customs (HMRC) and huge fines.

Getting the basics correct in payroll has become more complex over the last decade or so, however. Payroll departments must now ensure they provide data to HMRC in real time each time they pay employees, that eligible employees are auto-enrolled into a pension, and keep staff up to date with an increased variety of benefits. If a business expands globally, they need to get to grips with international currency and tax issues, while domestically they may be dealing with the complexities of seasonal, contractor and freelance contracts. And, during the pandemic, payroll experts had to step up and become experts on furlough, ensuring there was tight integration of HR and pay data at a time when most organisations were in a state of flux. "It can feel like herding cats sometimes," adds Barnden. "You're dealing with lots of different data sources, they all need to be accurate, authorised, compliant with your terms and conditions, and compliant with legislation." The 'glue' in the middle is likely to be your central HR system, which gathers the data and creates a single source of the truth from entering new starter data to running payroll.

The past two years have also shone a light on pay – showing it to be not just part of a transaction between employers and employees for work done, but how salary factors into employees' wellbeing and ability to function well at work. A recent reward survey by the CIPD found that 68% of organisations felt employees' financial wellbeing had suffered due to the pandemic, although almost half of employers did not have a financial wellbeing policy in place.[14] Even before Covid struck, the proportion of employees in the UK living 'payday to payday' with no emergency savings was 40%, according to Willis Towers Watson[15], and living costs are expected to rise further in 2022 and beyond as inflation reaches levels not seen for 30 years[16]. Worrying about whether their salary will be able to cover living expenses or an unexpected bill can have a negative impact on productivity, explains David Williams, head of group risk at Towergate Health and Protection. "If someone has a big bill coming up, their mind can go off the job, there are mistakes in emails or the production line. If more people become distracted things start to grind to a halt," he says.

One of the key factors in boosting employees' confidence in their finances is making sure they know what's available to them, not just in terms of salary but wider benefits and nonfinancial incentives, too. Employees who pay into a pension, for instance, may struggle to visualise the funds they need for retirement

14 Reward management survey, CIPD, March 2021, https://bit.ly/3qoCct1. Accessed 16/03/2022

15 40% of UK employees live payday-to-payday and half overspend each month, Willis Towers Watson, February 2020, https://bit.ly/3nnSkce. Accessed 16/03/2022

16 UK inflation climbs to 30-year high of 5.5%, Financial Times, February 2022, https://on.ft.com/3icVPzn. Accessed 16/03/2022

when that is decades away. Many organisations take a 'total reward' approach to how they offer and communicate benefits, where they offer a 'menu' of options each year – such as gym membership or cycle-to-work schemes – and employees can view a statement of what they receive. This visibility is crucial in terms of communicating the value of benefits beyond net pay, adds Williams. "There's a perception that with insurance, for example, the only time you get it is when you're dead," he says. "But many suppliers offer extra benefits such as bereavement support or an employee assistance programme. In the current labour market, employees are starting to take in what wider benefits are on offer and once they know what they've got, they're less likely to walk away for a bit more salary."

For HR and payroll teams, however, all of this adds to the growing complexity of the role and the underlying technology that supports it. "This is not just payroll systems but other systems that form part of HR data workflows such as expenses and benefits platforms," adds Barnden. Jeff Fox, principal at Aon, says this is because there has been a gradual broadening out of what constitutes reward. "We are seeing a minor revolution in how pay and reward is defined. It is more complex but potentially more meaningful," he explains. "This could be recognition, wellness, sustainability – the list is long. It challenges the notion of the original definition and makes the reward manager or director's role more holistic. To communicate the true value of what it means to work at a particular employer, pay and reward teams need to consider elements that are potentially part of the wider employee value proposition." And with the growing number of apps and systems available in the pay and reward space to help them do this, it's even more crucial that they integrate tightly with each other to ensure accuracy.

And, as employers in a number of sectors face skills shortages, this 'employment deal' becomes increasingly important for attraction, he adds. Reward has become a central part of what it means to work at a particular company, so communications need to be both more transparent and more coordinated. "The employment deal is no longer just pay – it now constitutes an arrangement where the employee can be part of something that makes a difference," adds Fox. "To avoid fragmentation in the proposition, a reward approach that is integrated can coordinate key messaging, organise communications and deploy relevant technologies to broadcast a clear and joined up story of the employment deal. The current age is defined by volatility – and total reward can take a lead role in helping manage this complexity," he says. Jeppe Rindom, CEO and co-founder of expense management platform Pleo, sees this taking the form of more autonomy for employees over their entitlements – for example, the ability to invest in equipment for home working without a drawn-out approval process. He says: "If we can foster this kind of autonomy and transparency as the default, we'll see fewer employees worrying about where they stand or having to dip into their own pockets for work expenses, and therefore far lower levels of stress within the workforce."

Being able to access pay more flexibly, rather than waiting until the end of the month, is one option for reducing employee stress. This is reflected in the growing use of early wage access apps such as Wagestream, which allow employees to access a proportion of their earned salary before their usual pay date. Employers can set a ceiling percentage for how much staff can access and can also set controls around how often cash can be accessed early. The apps link to companies' central HR and payroll systems, which can show how many hours have been worked, set how much the employee can draw out early, and 'balance the books' in time for the monthly payroll. "Early access to earned wages can stop people getting into debt during pay cycles," says Peter Briffett, CEO of Wagestream. "There's a dichotomy that if you're a low-paid worker you can be seen as high risk by credit card companies, so financial products can end up costing you more than someone who earns a lot." A study by the company found that stress decreases for 77% of employees who have access to flexible pay options, and budgeting improves for 55%.

If employers want to fully support staff through earned wage access, it's crucial to offer this as part of broader financial support and education. Looking at usage data anonymously can show trends among particular workers or teams, too, says Ian Hogg, CEO and co-founder of fastPAYE, which offers early wage access to hourly paid workers. "If someone always takes the maximum they can every month, that could highlight a potential financial wellbeing issue," he says. A number of early wage access tools have links to debt advice charities where employers can fund counselling or access to government benefits calculators, so workers can check if they or a family member could be entitled to state support. "The aim is to spark a conversation, so an employee could talk to a manager about getting more hours or be passed on to an external professional if it's a difficult issue. Access to money advice articles, budget advice calculators and savings calculators are also useful, although often the people who end up using these tend to be financially savvy already and therefore aren't the ones who need this support the most."

Reflecting this, the CIPD's reward survey found that just over a third (35%) of employers that were making changes to benefits in the past year had adapted their reward strategies to include financial wellbeing. This included alerting employees to financial scams, financial allowances for staff working from home, and early pay access. "This reward strategy also supports an employer's mission, vision and values, which come together to create the workforce's resilience and agility," the research explains. But it also sounds a note of caution: "The challenge is that making quick amendments to employee benefits in response to a sudden upheaval in the business environment can potentially weaken workplace culture if these amendments are not appropriately designed or implemented – in which case employers may be put off from making negative changes (such as a benefit cut) unless there is no alternative."

One of the benefits of early wage access systems is they can show a visible link between wages earned and the work itself, as most tend to offer real-time earnings tracking via an app on employees' phones. Briffet adds: "This is highly valuable because employees have not typically had access to earnings information in the palm of their hands before. It means they can start making decisions. If they're likely to go overdrawn they can do an additional shift or decide not to spend that money." Rindom from Pleo agrees – if employees can avoid having to refer to outdated paper-based processes, contacting shift managers or logging in to multiple systems, this can ease stress and allow them to concentrate on their work. "Employees can track their spending in real time, without worrying about all the hoops they'll need to jump through to get reimbursements they need for their own financial commitments," he adds. "It comes down to putting the employee first by empowering their spending experience."

In the longer term, the engines powering pay and reward in the background will become ever more sophisticated and subject to automation, says Barnden. "The big advances will be oriented towards automation and bulk processing," she predicts. "This won't be obvious to employees because they just want to get paid, but we'll see huge savings in terms of time, money and reduced variation."

Five key takeaways

- Remember that the 'employment deal' is more than salary: it covers all aspects of benefits and how they align to company values
- Payroll has become increasingly complex in the past decade or so, meaning systems integration and data cleanliness are crucial
- Consider early wage access for employees as a way to reduce stress about unexpected expenses

- Use data from pay and reward systems to inform how to tailor benefits communications and invest in new offerings
- Financial wellbeing has become a central part of organisations' overall wellbeing strategies, so ensure employees have access to resources

PART TWO

HIRING

Social media for hiring

Platforms such as Facebook, Twitter and Instagram are a constant feature in our lives – we spend (often too much) time refreshing our feeds or updating our statuses, satisfying our curiosity about what's happening in others' lives. The average internet user now spends a staggering six hours and 43 minutes online each day, with more than a third of that spent on social media.[17] The dopamine hit users get every time they receive a notification makes social media tantalisingly addictive. Not surprisingly, it makes sense for employers to build a brand and reach out to potential candidates via these platforms. Consequently 'social hiring', as it is sometimes known, has evolved as the platforms themselves have grown. According to employee review and job site Glassdoor, 79% of candidates are likely to use social media in their job search.[18]

The ways organisations can use social media channels to reach out to both active and passive candidates are as varied as the ways we might use them as consumers. Music streaming service Spotify has been known to use playlists to engage candidates (one recruiter at the company even used a playlist to make a job offer),[19] while retailers might ask shopfloor workers to make a short video of them in their role and post

---

17 Digital 2020: 3.8 billion people use social media, We Are Social, January 2020, https://bit.ly/3I2kmSH. Accessed 16/03/2022

18 Why investing in employer brand pays off, Glassdoor, January 2020, https://bit.ly/3fjjGvQ. Accessed 16/03/2022

19 This Spotify Recruiter Turned a Job Offer Into a Playlist—and the Candidate Loved It, LinkedIn Talent Blog, October 2018, https://bit.ly/3GxNfGc. Accessed 16/3/2022

it to Instagram. It can build brand awareness, and link back directly to a recruitment hub or application page on a company website, or even incorporate games or assessments that will filter the best candidates. The real-time data recording views, shares, click-throughs and swipe-ups can provide valuable insights to recruiters so they can adapt campaigns quickly.

Before embarking on a social hiring campaign, it's essential to consider your audience, says Alex Fourlis, managing director of Broadbean Technology. "For every geography and type of job there is a different media mix that each recruiter should consider," he advises. "Having a balanced mix of job opportunities, employer brand stories and useful info for jobseekers is critical when incorporating a social recruitment strategy." Peter Linas, EVP of corporate development and international at staffing software company Bullhorn, agrees: "Nowadays, posting job ads on social media allows companies to do more than just hire people. It enables them to increase branding, reach and engagement with talent who have already shown an interest in their company. It allows you to build a persona and is also powerful for referrals."

The big four

The platform(s) you choose should be guided by where you expect your candidates to be, which will be influenced by factors such as age, industry sector and type of role. Most of us access social media via smartphones, so content needs to be mobile-friendly both in terms of presentation and brevity. New social media platforms emerge all the time and it can be a challenge to keep up, but here's how to succeed with the 'big four'.

LinkedIn

The main advantage of LinkedIn is that it is purpose-built to connect professionals, so your job post or career message won't be bogged down by off-topic posts and stories. The platform offers several paid-for services, including LinkedIn Premium (which allows you to see profiles of those who are not connections and send messages) and LinkedIn Talent – a suite of services such as data analytics and recruitment marketing. But there are also a host of things hiring teams can do using the basic membership.

Joining relevant LinkedIn groups (industry- or location-specific ones, for example) will provide insight on the types of skills in demand and some will allow you to post links to job adverts for free. It costs nothing to look at similar job postings and benchmark them in terms of salary and desired skills or experience, or even to look at the profiles of professionals that are in those roles already. You could create a company LinkedIn page dedicated to careers, using videos and stories from employees to bring the brand to life; job posts on this page will also be free (although you may have to pay to create the careers area). Asking existing employees to share posts amplifies your reach far and wide and makes the most of their networks.

Paid-for job advertising on LinkedIn can be cost-effective. It runs on a pay-per-click model so you only pay when a potential candidate views your listing, and you can set daily budget limits or the maximum you'll pay for every 1,000 impressions. Marja Verbon, chief executive of careers site Jump, says that while it is often the first port of call among jobseekers, LinkedIn has its limitations. "LinkedIn allows access to thousands of professionals and as such can be a great tool for headhunting. But its outreach is very manual and conversion rates are very low, so it is not very scalable as a recruitment tool."

Facebook

The core attraction of recruiting through Facebook is its sheer size. At the end of December 2021, there were 2.91 billion monthly active users[20], so arguably more people will come across your vacancies on this platform than on any other. Facebook has

20 Meta reports fourth quarter and full year 2021 results, Meta, February 2022, https://bit.ly/3IfB8xi. Accessed 16/3/2022

its own powerful built-in search engine, Graph Search, which allows you to search against publicly listed profiles, meaning you can pinpoint people in certain locations and even roles or companies (if they've made that information available).

One of the most popular recruitment tactics on Facebook is to set up a company page to share photos and videos of employees to reflect a more three-dimensional view of your organisation's culture. Barclays, for example, has a dedicated Facebook page for early careers and graduate recruitment, showcasing stories from people who have been through its internships, apprenticeships, and graduate schemes alongside links where candidates can apply. As with LinkedIn, Facebook groups can be another cost-effective option to get your employer brand out there: student or alumni groups are a good option, as are specific-interest groups related to the role or your industry.

You can of course post a job ad on Facebook on your personal timeline, but this will be limited to your friends or friends of employees. Paid-for posts offer the option to boost and target your audience depending on the criteria you're looking for – using a general 'we are hiring' ad is a more cost-effective option if there are multiple roles, but for hard-to-source skills a specialised post could attract more attention.

Twitter

With limited word counts in each post, early social recruiters used Twitter as a way to simply post links to jobs, but it has evolved to become an effective way to entice potential candidates in to see more of your brand. For one, it's easier to start a conversation with your audience because there is no barrier of being 'connected' to someone – you can simply follow relevant talent and reach out to them. However, getting to know potential candidates before sending direct messages or tagging them in tweets about jobs is a good idea: what content do they publish or retweet? Who do they follow and does this reflect their professional interests? Are there people they interact with that could also be worth targeting?

A subtler strategy is to use Twitter to amplify your employer brand and recruitment marketing efforts and show off the company's personality. The character count allows you to embed links to videos or longer blog articles that showcase your culture. Social media scheduling company Hootsuite does this well: its Hootsuite Careers handle (@hootsuitelife) includes a range of stories from how the company has responded to the Black Lives Matter movement to how it supported an employee to relocate. For cost-conscious recruiters, perhaps the main advantage of using Twitter for recruitment is that it is completely free, although paid-for advertising options are of course available.

Instagram

The picture and video-led platform is often favoured by younger audiences so is good for showcasing workplace culture, or using imagery to spark a conversation with potential candidates. There are around 1 billion monthly active users on Instagram, and 70% of them are under 35.[21]

Claire Stapley, a recruitment marketer who has worked on several hiring campaigns, says that content has to add value to the target audience if it is to be effective. "This shows you're credible and know your demographic. Members of generation Z [born between the late 1990s and early 2010s], for example, are thinking [more] about inclusion and an organisation's impact on the environment and are less concerned about salary," she says. "It's not about sharing how great your company is, but outlining initiatives you may have, content you've written or something relevant from another platform. As long as it's in line with your brand values, that's great."

The type of content you could share on Instagram can be broken into three key areas: people, events, and environment. People stories might include celebrating an employee's achievement, and

21 Distribution of Instagram users worldwide as of January 2022, by age group, https://bit.ly/3rhPuqp. Accessed 16/3/2022

event posts could showcase the social side of the business, while environment-themed content can give candidates a feel for what the workplace itself will be like. Choosing the right hashtags is critical on this platform – you could create a dedicated hashtag for your brand as well as use popular ones such as #nowhiring, #jobs and #jobsearch. Cisco's dedicated @wearecisco account is a good example of life at the technology company; there are few direct references to open roles but a huge mix of stories and videos about its employees and ethos.

The new kids on the block

Don't limit your candidate search to the four key platforms. Potential applicants don't need to be actively conducting a job search for you to get your brand in front of them – think about other channels where they could be socialising online. Here are a few to consider:

TikTok: in 2019 this video sharing app surpassed 1.5 billion downloads and is hugely popular among generation Z audiences. The app revolves around sharing 15-second video clips so the brand message needs to be concise.

Snapchat: another popular platform among younger people, Snapchat allows users to share photos and videos for a finite amount of time. McDonald's ran a 'Snaplications' recruitment drive, with 10-second videos of employees discussing what it's like to work there, with a swipe up to its recruitment page.

Pinterest: not just for checking out bathroom tiles and recipes, picture-led platform Pinterest allows users to 'pin' posts they like and recruiters can access paid options to promote pins that are performing well.

GitHub: if you're looking for a software developer, you'll probably find them here. Developers use the platform to discuss projects they're working on. Profiles and locations are visible and you can search for specific programming languages.

To boost the candidate experience, you can offer integrated application routes in the posts (both LinkedIn and Facebook offer 'apply now' options) and deploy chatbots so candidates can ask questions at any time. Combining data from social channels and other sources such as your careers site and jobs boards can be challenging but insightful, says Linas. "Analytics is tough to do on your own, so you need the right technology stack that enables you to pull relevant data. Doing this will enable you to see where candidates are coming from, and linking this data with other sources further on in the employment lifecycle (such as engagement, retention and performance data in your HR system) could provide insight into which sources produce the most productive and successful applicants."

It's worth remembering, however, that social media can show off your brand reputation in a less flattering light, too. Glassdoor, where employees can leave reviews about current and former employers, has become a go-to research point for candidates. Because the reviews are generated by employees, potential recruits trust them as a 'warts and all' reflection of an organisation. "Employee-generated content such as reviews can be excellent recruitment marketing content and be used across multiple channels, including social media," says Joe Wiggins, Glassdoor's director of corporate communications. The power of employee reviews is in their transparency, so building and showing off an authentic and fair company culture is key to creating an audience of engaged candidates. In 2020, the site introduced specific diversity and inclusion ratings, offering potential recruits another benchmark on which to base their job search.

Anyone who has watched the documentary *The Social Dilemma* will know that the public are becoming more cautious of social media, with users beginning to reconsider and re-evaluate their relationships with these networks. Whether this trend will last and what its impact might be on how hiring teams engage with candidates via these channels remains to be seen. But for now, employers cannot ignore the potential of reaching out to a global network of talent at comparatively little cost.

Case study: Pension Protection Fund (PPF)

The PPF's social media strategy focuses heavily on recruitment, helping it to attract, recruit and retain a diverse employee population. The organisation, which protects members of defined benefit schemes if the schemes they've paid into fail, uses LinkedIn, Twitter, Instagram, and Glassdoor, along with job adverts, to recruit new hires.

One of the reasons PPF uses social media is to reflect its diverse workforce and culture, and to support this it uses searchable hashtags such as #inclusive, #diverseemployer and #disabilityleader. "A key feature of our strategy is the use of video to give candidates a flavour of what working for the PPF is really like," says Glenda Kladitis, PPF senior HR business partner.

"For instance, when advertising a new role we ask the hiring manager to record a short video talking about the team, opportunities and life at the PPF. This adds a human element and brings the role to life, giving the candidate another dimension they couldn't get from reading an ad."

Each of the social channels serves a specific purpose, and LinkedIn and Glassdoor are the core routes to candidates. "We are fortunate to receive enough applications organically that we don't currently need to use LinkedIn's paid features," she says. Instagram is aimed at current and prospective employees, giving a 'behind the scenes' view of life at the PPF, showing, for example, candidates discussing how reasonable adjustments were made for their disability during the interview or onboarding process.

Most applications to the PPF come through social media, and this has largely eliminated the need to use agencies and headhunters. "We're also proud to hear anecdotal feedback [from people] who have been attracted by the diversity they see on our channels, which is a true representation of our workforce. We aim to be an employer of choice for people of all backgrounds and are pleased to see our efforts in this area being effective," Kladitis adds.

Five takeaways

- Almost four-fifths of candidates use social media in their job search, so having some form of social media presence is essential. If you have limited experience with social media, why not ask your organisation's marketing team for advice?
- Social hiring can complement or even replace paid-for job advertisements, making it a cost-effective route to candidates
- Visual channels such as Instagram and Facebook can provide a realistic view of what it's like to work at your company
- The potential reach of social media means it can project your employer message far and wide, so make sure that's a positive one
- Social hiring campaigns produce rich data and insights on where candidates come from, which can be used to tailor future recruitment efforts

Integrating your social media data

Integrating these campaigns with your recruitment platform and ATS can make them even more effective. Find a plug-in tool that will feed social media data into your ATS and this can maximise your efforts. Tools such as Broadbean can do the following:

- Automate posting on chosen social media platforms, reducing time spent doing this manually
- Post on multiple platforms at a time and manage advertising spend
- Track referrals and reward employees for sharing posts
- Upload candidate information into your ATS to create reports on conversion rates so you can tweak campaigns
- Create a searchable database of candidates so you can filter against selected criteria
- Establish a record of candidate information, making onboarding quicker if they are successful in securing a role

Matching candidates to roles

Unless you have the luxury of recruiting for a niche role that has attracted just a few, highly qualified candidates, coming up with a shortlist can be a time-consuming task. Recruiters in organisations undertaking high-volume hiring campaigns – for example, cohorts of graduates or staffing a new contact centre facility – face a mountain of potentially thousands of applicants, and unsurprisingly have turned to technology to help them sort the good candidates from the unsuitable ones.

Candidate matching has become increasingly sophisticated in recent years and is one of the most common applications of AI in HR and recruitment. Dozens of software companies have emerged to sell tools for every step in the hiring process – from tech that matches job postings with likely candidates to tools that scan applications. There are also tools that can pick up keywords in CVs or applications, and those that 'score' candidates against a set of criteria for the role or deselect those who don't have the required qualifications. All claim to do so in a matter of minutes compared to days working through applications manually.

Providers of such systems claim they can significantly reduce recruiter workload, improve time to hire and save organisations money because technology can automate much of the sifting process. "While human judgement will likely remain more reliable, there are parts of the hiring process that can be automated. For hiring managers, the biggest bottleneck is having to sift through endless unsuitable CVs," says Jump co-founder Verbon. Its preselection process checks candidates against a list of criteria before they apply. Only an estimated

12% meet those criteria, meaning screening time is potentially reduced by as much as 88%.

Sound too good to be true? AI's growing role in matching candidates to jobs does have its limitations. At a basic level, most algorithms will search based on keywords in CVs – so hiring managers could be missing out on the bigger picture of what a candidate has to offer. "Matching has got a lot better," explains Neil Armstrong, commercial director at recruitment and onboarding software company Tribepad. "But a CV is not necessarily a good way of understanding someone's capabilities. You can get a view of someone's experience but no insight into their personality or potential. And people have learnt to game the system, by putting in certain buzzwords that an AI search will pick up. They get found out later in the process because they don't have the skills, and it's a waste of time for them and the recruiter."

One of the key concerns around the use of AI in candidate screening is its potential to introduce or reinforce bias. While on the one hand it can reduce humans' unconscious biases by automating the sifting process, the algorithms on which this automation is based can in themselves be problematic. Often, employers are using historical datasets against which to score applicants, meaning they could be stacking the odds against underrepresented groups including women or black, Asian and minority ethnic candidates. Furthermore, certain keywords may work against certain groups. For example, if you ask the algorithm to identify applicants who play a certain sport or have a particular behavioural attribute, this could inadvertently exclude a sizable group of suitable candidates. Outside of recruitment, biased algorithms used by the US courts have been shown to mistakenly label black defendants as having twice the potential to reoffend as their white counterparts.[22]

Kim Nilsson, co-founder of Pivigo, a data science recruitment specialist, explains: "You would hope that as [algorithms] get

22 Machine Bias, ProPublica, May 2016, https://bit.ly/3fnfvz6. Accessed 16/03/2022

more sophisticated that they will remove bias, rather than add to it, but the slightly scarier question is: as algorithms get more embedded or prevalent, how big a risk is bias to the process?" This is because algorithms learn from 'training data' based on past success, she adds. "In HR, a training dataset would be, for example, a set of CVs from previous [candidate] applications with labels of which ones had successful job offers. The problem here is that if you feed this sort of data to the algorithm, it may say that someone with a different profile from your current workforce (such as gender, nationality, or educational background) should be rejected because your previous 'success' cases do not have this diversity in it. You will perpetuate the biases that already exist in your workforce and data."

"Greater diversity in algorithm development teams could help mitigate unintentional bias into system design," says Nimmi Patel, policy manager for skills, talent and diversity at UK industry body techUK. "Having diverse teams – in terms of gender, ethnicity, experience, and background – will increase the likelihood of unconscious biases being recognised and addressed, rather than encoded into future algorithmic-decision making systems. This will, in turn, improve the quality of the decisions being made."

Gareth Jones, CEO of Headstart, an AI system for graduate and early career recruitment, agrees that the data going into the system needs to be improved if AI is to deliver on its promises. "AI offers the opportunity to make hiring fairer, but we're still a long way from hiring being data-driven," he says. "Recruiters still end up looking at a CV or a profile. They may supplement that with assessment, but it's not driven by what the success criteria are in that company. That should be the starting point – not a job description."

That said, there are ways to make the most of AI by combining it with other tools to predict good hiring outcomes more accurately. Introducing assessment into the process can be a means of building a more rounded and objective view of the candidate. "These can be integrated into a company's ATS at the start of the hiring process, after the initial candidate application," explains Chris Platts, CEO and co-founder of ThriveMap, a pre-hire assessment company. ThriveMap's assessments take candidates through a digital 'day in the life' experience of a job. "Candidates are automatically invited to complete their 'virtual shift' via email or text message and the results and interview reports are sent directly into the ATS," he adds. "This real-world approach to talent assessment means that candidates can get a feel for the role and culture ahead of joining the company; they can even opt-out of the hiring process if they feel it's not right for them."

Candidates deselecting themselves from the process is actually a positive outcome for hiring managers. It ensures those moving to the next stage identify with the organisation's culture and are more likely to be productive and engaged if they're successful in getting the job. Data and feedback provided by assessments during the hiring process, furthermore, can help demonstrate to eager but unsuccessful candidates why they did not meet all the criteria and how they could boost their skills when applying for another role – or even suggest a more suitable one at the same organisation. "Context-specific tests can showcase the role is real. So if you're looking at a PA role, here's an enquiry we'd like you to deal with, rather than asking for generic skills such as Excel," adds Adrian McDonagh, co-founder of hireful, a company that supports recruiters to improve their hiring practices. "Then you can point to the assessment and show why you made the decision rather than just saying 'we found a better candidate'."

Richer assessments can also show recruiters applicants with a diversity of values and thinking styles. A 2013 report[23] by Deloitte recommended organisations hire with 'diversity of thought' in mind in order to protect against groupthink and promote new ways of solving problems. One way of doing this

23 Diversity's new frontier: Diversity of thought and the future of the workforce, Deloitte, 2013, https://bit.ly/3fmcHSW. Accessed 16/3/2022

is through game-based recruitment tasks or challenges, which can show not only whether candidates have job-related skills but also how they reach decisions – an important consideration when looking to build a diverse team. British intelligence agency GCHQ has run multiple campaigns – aimed at hiring people with cyber skills – which have challenged candidates to crack codes or decipher messages, in line with its requirement for applicants to 'think like a hacker'.

Employers such as Lloyds Banking Group and Police Now, meanwhile, have used virtual reality challenges for graduate recruitment where candidates complete simulations and recruiters can see how they navigate them. Some game-based hiring challenges go as far as to offer a prize: Google's Code Jam, where programmers compete to show off their coding skills, offers a reward of $15,000 for the winner (and a solid shortlist of candidates with proven coding skills for Google).

Knowing more about how the candidate operates in a realistic environment and whether they might make a good fit with the team makes the onboarding and settling-in process a lot smoother, adds Platts. "Identifying candidates in your applicant pool who align with your company's desired behaviours, who can demonstrate the required role capabilities, and have a genuine commitment to want to do the job is difficult," he says. "Making better hiring decisions means that candidates become productive more quickly. If assessments take candidates through a digital experience of the job, they're less likely to be surprised by what the job involves when they start and leave prematurely." Effective use of algorithms can mean organisations can better target their recruitment marketing.

Verbon adds: "On the professional side, job recommendations up until now have been extremely hit or miss. We've all had our fair share of utterly irrelevant job alert emails. What algorithms can help with is identifying which jobs would be a good match based on the professional's CV, rather than just random keywords. These recommendations are much more accurate and can make job hunting a far less daunting task."

Ultimately though, AI and other automation tools should be used to support and augment what the human recruitment team does, rather than to replace it. "Used correctly, these systems can do wonders for improving diversity in an organisation, by promoting the applications of individuals who otherwise would not come through [the shortlisting process]," says Nilsson. "HR individuals will often scan hundreds of applications and will for obvious reasons not have time to study each application in detail." They can also help to overcome the shortcomings of mechanisms such as referrals. "Recommendations may be a good indication that that individual could be a good hire, but it will miss out on stand-out talent who are not as well networked. So, algorithms can support the decision by highlighting applicants that would otherwise be hidden in the noise, without actually making the final decision," she adds.

A new start-up called BrightHire, backed by renowned organisational psychologist Adam Grant, offers to bridge this gap. A professor at the Wharton School of Business, Grant's book *Originals: How Non-Conformists Move the World*, looks at why – in the workplace – it's not always conventional qualities or actions that make people good at their job. BrightHire's tool, rather than simply automating decisions about who makes the cut, provides context to 'flesh-and-blood' interviewers during and after video calls with candidates. The software offers an 'interview assistant' that keeps interviewers on track, displaying the predetermined questions they'll need to ask each candidate. By asking a standard set of questions, the interviewer is less likely to make subjective evaluations based on their own biases, while the process is rendered more efficient through BrightHire's prompts. Pre-hire assessments are becoming more sophisticated, too, including neuroscience-based tools that can predict whether a candidate is likely to be a good cultural and skills fit.

Faced with an overwhelming volume of applicants, the promise of speeding up the matching process is an attractive one — whether at the start of the candidate search or when you're inviting people to interview. A dizzying array of technology is available to support HR and recruitment teams in finding, matching and shortlisting candidates – and while this can't replace the human touch, it can make the process more evidence-based, efficient, and cost-effective.

Five key takeaways

- Recruiters can now choose from a huge selection of tools that can support candidate matching and selection
- AI and algorithms can save time and money for hiring managers, but should augment human decisions rather than replace them
- Be aware of the potential to build bias into recruitment algorithms
- Technology can build a more reliable picture of a candidate's suitability, making them more likely to be a worthwhile hire
- Establishing someone's suitability for the day-to-day aspects of a role in the hiring stage can boost retention and productivity

Using video in recruitment

The coronavirus pandemic prompted a vast range of changes to HR processes, but one of the most dramatic shifts was the sudden shift to video recruitment. With restrictions on in-person meetings in place through much of 2020 and into 2021, interviews or briefings from hiring managers often needed to happen over video communications tools such as Zoom or Microsoft Teams. Hundreds of successful candidates started roles without ever meeting their manager or team in person – and, with hybrid working here to stay, it's likely that thousands more employees will follow in their footsteps in the coming months and years.

The necessity to shift much of the recruitment process to video has debunked many of the myths around it being a lesser way to engage with candidates. According to a 2020 survey of candidates by RecRight[24], 82% are comfortable with video interviews as a method of recruiting. For the companies doing the hiring, the benefits have been magnified – in particular, the reduction in time to hire. According to Kevin Parker, CEO of video hiring platform HireVue, US supermarket giant Walmart reduced its hiring process from 14 days to just three by using video interviews, and was conducting as many as 15,000 a day when Covid-19 restrictions began.[25]

James McGill, vice president for customer success at HireVue, explains: "Our data shows a large acceleration of remote interviewing across the last 12 months with the ongoing

---

24 RecRight Candidate Survey, 2020, https://bit.ly/3I5qObM. Accessed 16/03/2022

25 One retailer is hiring amid Covid-19 and hiring fast. Here's how, My Total Retail, April 2020, https://bit.ly/3ttTVBh. Accessed 16/03/2022

coronavirus situation leading many organisations to update and adapt their recruitment processes. But adoption of the technology had already been increasing for a number of years, and was on a steep upwards trajectory, even before the impact of Covid was felt, due to the cost-saving and streamlining effect of virtual hiring solutions."

He adds: "The introduction of on-demand and live video interviewing can greatly reduce many of the time constraints and costs associated with the traditional recruitment process, offering greater flexibility and time savings to both candidates and recruiters, allowing companies to access the best candidates without the limitations associated with on-site interviews." Furthermore, McGill notes the HireVue platform, for instance, enables multiple members of a hiring team to review a playback of a candidate's interview in their own time without having to sync calendars, and 'live' interviews can happen in spite of any social distancing restrictions or location clashes.

Video interviews can be conducted in two ways: either in real-time, replicating an in-person interview over a platform such as Microsoft Teams or Zoom, or asynchronously as McGill suggests. Asynchronous interviews had been growing in popularity even before the pandemic, particularly for volume recruitment where they can be used as an initial screening stage. Another growing trend is to ask candidates to send in a short video alongside their CV or application. This gives a richer idea of someone's personality and whether they might be a good cultural addition to the organisation, or how they might perform in the role on a day-to-day basis – particularly useful in customer-facing roles. Most video recruitment tools include some level of AI, so can scan hundreds of videos for keywords or phrases, shaving off hours of interview time that would have been spent on in-person interviews.

Video can be used at multiple stages of the hiring process: to be incorporated in the initial communication with shortlisted candidates (a video introduction of the hiring manager and/ or company); a video of the questions you wish to ask the candidate to which they respond; and the candidates' answers. McDonagh from hireful has seen more and more hiring managers record responses to three or four common questions about the role or department. "As the pandemic has made people more comfortable with video, we might ask the manager some questions to share with candidates – it makes good sense and people feel well-briefed," he says. Embedding video into the job advertising itself is great for candidate attraction; a Cisco study found that recruitment agencies reported 800% more engagement with job postings that had videos in them.[26]

Recruiters recommend that asynchronous video interviews (particularly in volume hiring) should work on the 'less is more' principle. There should be limited questions (three to five is ideal) and hiring managers should be very clear about the information they need from candidates and the types of qualities they will be looking for. Candidates should be instructed on the length of their answers and given any other specific guidelines. When it comes to live interviews, it's crucial to upskill hiring managers in the skills they might need to handle a virtual conversation – such as waiting for an interviewee to respond before cutting in, being mindful of potential connection issues, and making allowances for interruptions.

There are numerous benefits to offering candidates the chance to record interview responses, including:

- Offering the same set of questions to each candidate means the process is fair and consistent
- It widens the potential talent pool because applicants that might struggle to get to an interview can 'join' at any time
- It reduces the need to physically bring together hiring managers, who may have other commitments. They can review selected responses in their own time

26 Cisco Annual Internet Report (2018–2023), updated March 2020, https://bit.ly/3r9p2za. Accessed 16/3/2022

- Automating the video review process reduces the potential for unconscious human biases to creep in

Because many video recruitment tools are based on AI, one of the key advantages of their use is the opportunity to simultaneously reduce screening time and the tendency towards unconscious bias. "Video interviews can help mitigate against the risk of unconscious biases held by recruiters, offering results based solely on the data shown to predict job success," adds McGill from HireVue. There is even a growing market for pre-screening tools, also powered by AI, that use facial recognition to assess whether someone is suitable for a job. These tools measure characteristics such as whether someone comes across as confident, calm, excited or enthusiastic. They can even detect small movements of the eyes, mouth or jaw that could suggest the candidate is lying.

But, as with the prospect of asking candidates to provide a video application or 'showreel', this approach is not without risk. If an algorithm judging facial expressions or gestures is based on biased data, this could mean it's only picking out those who conform to certain expectations of what 'confidence' or 'enthusiasm' look like, and could even be potentially discriminatory against people with neurodiverse conditions such as autism whose expressions don't always match what they're trying to say.

To get a broader range of evidence to support a hiring decision, recruiters can use video alongside other tools such as competency assessments or psychometric tests. Data or results from this stage of the process can then be linked back to the central HR system to provide insights such as learning priorities or where there are gaps in candidates' knowledge. "Using assessments, for example, you could send a candidate with a good score a task to showcase what they can do over video based on a job specification," adds McDonagh. McGill argues this is especially useful when recruiting in high volumes. "The use of interview assessments, particularly in the early stages of the recruitment process, can help companies to accurately and optimally assess the competency of a huge volume of candidates. This is crucial for any business that has urgent vacancies, receives mass applications, or is operating in an industry that requires hiring at scale." Additional features such as game-based challenges mean hiring managers have "information at their fingertips so they can base their screening and hiring decisions on an objective, bias-free standard," he says.

Many of the benefits of asynchronous video interviews can also be found in live video interviewing – which Armstrong from Tribepad believes will now feature prominently in hiring processes. "Traditionally organisations have used asynchronous video interviews for that initial screening element, but they're also carrying out live interviews now because they can't meet face to face, with real benefits," he explains. "They're easier to attend, interviews can be done much more quickly, and if you have a good candidate that's considering other offers you can get to them quickly, making you a more attractive employer." Recording interviews means managers can see a transcript even if they weren't present, and this can be translated into other languages or be combined with assessment results and scoring tools to gauge candidates' suitability.

Despite the richness of video, hiring managers need to be wary of adding too many elements into the screening process. "People are cramming so much into the application process – asking candidates to do a video, engage with a chatbot, complete an assessment," says Jones from Headstart. "If a candidate has to do that multiple times it becomes repetitive, so it's better to invest in improving the process and asking why you need the video rather than just adding it for the sake of it."

That said, once a candidate ultimately moves to the second or third interview stage, and hopefully a face-to-face meeting with key stakeholders, there should be a clearer understanding on both sides if they are a good fit – and the candidate will likely be more invested in the process. McDonagh says: "Coming into offices creates a lot of wastage in terms of time, and video means organisations can move a lot faster in hiring for the role and there's more flexibility for the candidate. So, once they do meet the hiring manager in person, they know they've been properly shortlisted."

Given the growing popularity of video recruitment even before remote and hybrid working became more widely adopted, it will likely remain a key element of the hiring process for many organisations in the future – but should we try to replace the element of human interaction in recruitment if hiring over video is so efficient? In his 2019 book *Talking to Strangers: What we should know about the people we don't know*, Malcolm Gladwell discusses how someone's first impressions based on a half-hour job interview can often be wrong – that all we see is what someone looks like and how they handle themselves, with no evidence of their ability to do the tasks required for the role or to predict their long-term contribution to the organisation. Conversely, neuro-economist Paul Zak argues that we build trust through human contact.[27] And many would argue that for certain roles, at the final selection stages at least, it is non-negotiable not to meet with a candidate.

What is certain is that, with a growing range of remote hiring tools on the market and the algorithms behind them becoming more sophisticated, video will have an increasingly important role in augmenting – rather than replacing – face-to-face interaction.

Five key takeaways

- Video can dramatically reduce screening time in volume hire situations
- Hiring managers don't have to replicate live interviews over video: asynchronous interviews can offer multiple benefits

27 The neuroscience of trust, Harvard Business Review, January–February 2017, https://bit.ly/3qoPXbl. Accessed 16/03/2022

- Most video recruitment tools use AI to reduce the time spent reviewing candidates and mitigate human bias
- Think about adding assessments or other elements to enrich your view of the candidate – but don't overload them with hoops to jump through
- Video interviews are not just for pandemic recruiting: they'll be a key feature in the hiring process for years to come

Managing the candidate relationship

Jobseekers have never had more power. Before the internet revolutionised how organisations advertised roles and candidates could find a review of an employer with a few clicks of a mouse, employers would set an application process and everyone would have to follow it. Unless you were successful in getting the role, it was rare to know how many others applied or whether your skills might have been a good fit. Applications or assessments that might have taken hours of precious time were never to be seen again.

Since then, a glut of technology has emerged to enable companies to track candidates, build talent pools of potential applicants they've gleaned from past campaigns or other interactions, and to engage with those they want to bring on board. In a tight labour market, tools such as ATSs are a must-have for employers looking to reduce their time to hire and manage high volumes of applicants without overwhelming the recruitment team.

At the same time, however, candidates have come to expect an experience more akin to something they would get as a consumer, says Johnny Campbell, CEO of hiring consultancy SocialTalent. He references Matt Watkinson's book *The Ten Principles Behind Great Customer Experiences* as a guide that can equally be applied to the candidate journey. "Think about something as simple as being put on hold – the brand that tells you you're 26th in the queue is better at managing your expectations than the one playing hold music," he says. "It's all about the peak and the end – people will remember the peak of their experience and how it ended, and their experience is an aggregate of all these small actions. At each point your organisation has the opportunity to make it a better experience."

And while a great experience will make candidates feel good about your company because of how you treated them, a bad one will make them lose respect for you both as an employer and as a brand. According to Talent Board, which conducts research on candidate experience, the proportion of applicants willing to sever their relationship with a company after having a negative experience has increased by 25% since 2016.[28] LinkedIn's Talent Trends research, meanwhile, has shown that 94% of candidates want feedback after an interview, yet only 41% have ever received it.[29] Prospective recruits expect the process to be seamless – so whether they're just registering their interest in a career with your organisation, moving on to the application process, carrying out an assessment, or sharing a role with a friend, bouncing them around poorly integrated HR and recruitment systems means there are multiple opportunities for them to drop out.

With demands on recruitment teams to reduce time to hire, or to make sure they keep hold of in-demand candidates before they are snatched up by a competitor, it can be tempting to automate as many of these stages as possible. But Katrina Collier, founder of consultancy The Searchologist and author of *The Robot-Proof Recruiter*, advises against taking this approach. "If the tech doesn't put the human first, or save you time to put the human first, it won't work," she says. "Recruitment is complicated and involves opinions and emotions. You need to build trust, and technology can break that trust." The ubiquity of job boards, quick-apply buttons and social media means recruitment teams are often dealing with huge amounts of candidate data every day, but it's still possible to use technology

28 Talent Board 2019 EMEA Research Report, Talent Board

29 Talent Trends 2015, LinkedIn, 2015, https://bit.ly/3rcoitn. Accessed 16/03/2022

to filter through the noise and build positive relationships with potential hires.

An ATS can help you identify where candidates are in the process and keep up with compliance requirements, but is that enough? As a workflow tool and repository for candidate data, an ATS is sufficient, but arguably a candidate relationship management platform will perform better in giving you those 'richer' interactions with candidates that make them feel good about your brand – even if they don't ultimately get the role. Put simply, they are a system of engagement rather than a system of record.

With support from a robust candidate relationship management system, hiring teams can break down the stages of the process and curate which elements can be automated without breaking that all-important bond with potential hires. At each point, the message can be tailored to reflect the employer brand or the expectations that might be going through the applicant's mind at that stage. Crucially, they can also funnel information from multiple sources to build a talent pool for future campaigns, aggregating data on active candidates (who may have applied for a role before or have proactively contacted the company via a careers portal), as well as passive ones. This means recruiters don't have to reinvent the wheel with each new hiring campaign, saving time and money on recruitment marketing and attraction. A candidate relationship management system should help you engage with candidates who are not actively in the recruitment process or were unsuccessful the first time around – such as through targeted content or information about company developments – so they're the first people you contact when you're hiring for a similar role.

"The personal touch can be in the setup of the system," says Campbell. "So an automated response to a job application can be engineered to reflect the brand. You can empower candidates to set up their own video assessments and do them in their own time. You could set up an automated message they receive the night before an interview giving them tips on how to get to the location or how to access the video-conferencing software." Because applying for a new job can be an emotional experience, people tend to buy into the 'personality' of the brand and crave familiarity – so it's often about how you automate elements of the process rather than breaking things down to an 'either/or'. Proactively providing candidates with an insight into your company culture can also prompt those applicants who don't see themselves as a good fit to deselect themselves from the process, ensuring the ones who make it to interview or are eventually hired are genuinely bought in to what you do.

Here are some examples of how automation in the candidate engagement process can be humanised:

- An email acknowledging an application and signposting next steps, including setting expectations on when to expect a response
- Inviting candidates to book their own interview time through a self-service system rather than sending emails back and forth
- Responding to unsuccessful applicants (in early stages) but including career development content or even a voucher for your product or service
- A rich media text message that reflects your employer and consumer brand, which confirms you received an application and directs the applicant to your social media channels

Managing expectations and providing transparency around where candidates are in the process is as important as humanising your interactions with them, and your candidate engagement system should be able to support this. Telling someone they'll receive feedback and never giving it is unacceptable, says Collier, and this is where automation can help. "A tool to prompt hiring managers to give feedback is where we can show humanity in the automation of the details," she says. Key 'moments' in the journey – such as giving final interview feedback or making someone an offer – can be done face-to-face or over the phone, but that doesn't mean your underlying technology can't provide a useful prompt.

Another important consideration is whether candidates feel like they have to 'game' the system just to progress. For example, many candidate relationship management systems allow hiring teams to input rules so they can filter out the less suitable candidates, and progress others through to the next stage. "Think about the keywords you use or whether someone who doesn't have a degree might get a knockback," adds Collier. "This introduces bias and makes candidates feel as though they have to game the system to get through to a human or make it to the next stage."

The data generated by a candidate relationship management system can help hiring teams to get feedback on their own performance. Adding an anonymous survey can show whether candidates have valued the experience, explains McDonagh. "You can measure this from a diversity perspective too, to see if minority or outlier candidates have had a different experience – treat it like a mystery shop." Asking candidates for feedback before a selection decision is made – ensuring the questions are only valid for a certain time – helps to make sure this feedback is unbiased. Any ATS or candidate relationship management system should be able to produce data on simple metrics such as how long it took to respond to a first contact or the length of time between final interview and offer.

McDonagh believes many organisations could improve how they interact with candidates for a relatively low cost and effort. "It's the difference between making candidates feel like 'one of us' and beans on a shelf," he concludes. Setting goals for how you treat candidates in the same way you might for customers – for example, publishing a candidate charter on your careers page explaining all the things you will do (or not) during the hiring process – is a good start. Your candidate engagement technology is simply the engine that makes sure you keep those promises.

Five key takeaways

- Do personalise your candidate experience – no one wants to feel like just a number
- Automate simple processes such as interview arrangements, but humanise key moments like making an offer
- Use the data from your candidate relationship management technology to spot pain points and improve them
- Enrich automated responses with your organisation's personality, even if the candidate is unsuccessful
- Ask for candidate feedback on the process – and use it

Background and right-to-work checks

A 'white lie' on a CV or a candidate elaborating their experience in an application is not just a black mark against their honesty – it could also spell future trouble for your organisation. In 2018, an investigation by BBC Radio 4's *File on 4* found that NHS consultants and nurses had been buying bogus degree certificates from non-existent universities via 'diploma mills' in Pakistan.[30] In this case, the falsification of qualifications could seriously impact lives – but there are many other potential consequences of failing to do a thorough background screening of a potential candidate, including exposing your business to security breaches or serious crime.

During 2020, the Covid-19 pandemic increased pressures on HR teams to ensure they screen candidates adequately and avoid these risks. Jamie Allan, background checking consultant at Experian UK and Ireland, explains: "Sectors that did not have a significant requirement for background checking until recently now have significant volumes of work to get through. The increase in workload is particularly pronounced in sectors where there has been a rapid increase in vacancies, creating a labour shortage in that area of the market." Allan describes a twofold impact: some teams will struggle to keep up where it's hard to find talent or there are a lot of vacancies, while in other areas there may be ample work so candidates "reach for roles beyond their ability, experience or

30 'Staggering' trade in fake degrees revealed, BBC, January 2018, https://bbc.in/3I6TWiP. Accessed 16/03/2022

qualifications, and become tempted to fake aspects of their CV to reach the interview stage." The fact that more people have been working from home than usual also creates a security risk of people having access to sensitive data while off site.

There are multiple points during the recruitment process at which a candidate can provide information or documentation that is not an authentic reflection of their experience, qualifications or right-to-work in the UK. This can take many forms – from the relatively harmless act of a candidate elaborating their levels of experience on their CV to obtaining fake qualifications and documents. A study by recruitment site CV-Library, for example, found that more than 90% of applicants had lied on their CV, with two-thirds admitting to an attempt to make themselves look more experienced.[31] If this fraud is uncovered later in the employee lifecycle, it could mean your organisation is at risk of reputational damage, or worse – legal action. Former CEO of Yahoo!, Scott Thompson, was forced to step down after just four months in the role when an investor challenged his claim that he had a computer science degree.[32] While he had not been hired on the strength of his claimed degree, his public position as CEO meant he had a greater responsibility to be transparent about his employment record.

With hiring managers often facing large volumes of applications or CVs to wade through on top of their day job, it's perhaps not surprising that these embellishments are not always detected. Research by recruitment specialists Alexander Mann Solutions in 2019 found that 53% of recruiters do not detect fraudulent CVs until the interview stage, and 49% during the background checking phase.[33] Common falsifications included altering employment timelines, inflating job titles and listing fake

31 These are the 5 areas Brits are willing to lie about, CV-Library, April 2018, https://bit.ly/3K8PPEr. Accessed 16/03/2022

32 Yahoo confirms CEO is out after resume scandal, CNN, May 2014, https://cnn.it/3nojj7J. Accessed 16/03/2022

33 Risk Mitigation in Assessments: The silent threat of CV fraud, Alexander Mann Solutions, 2019, https://bit.ly/3fsks9L. Accessed 16/03/2022

qualifications. If someone has faked documents or references, they may also pose a higher risk of committing a crime within your business, says Tony Machin, CEO of verification technology supplier TrustID. "Internal crime is one of biggest concerns for businesses – but they don't like to talk about it," he says. "If someone has fake documents, you will struggle to find them." Furthermore, in certain roles, such as working with children or vulnerable people, where a Disclosure and Barring Service (DBS) check is required, an applicant providing a fake document could create a dangerous safeguarding issue. Having a system in place to keep an audit trail of checks and balances made during the hiring process is therefore crucial. This could be as simple as an Excel spreadsheet or something more sophisticated such as bespoke screening software or services, like TrustID's, that works alongside your organisation's applicant tracking and HR systems.

Automating aspects of the screening process can also reduce the potential for human error. The checks and balances offered by integrating screening capability into your HR system mean that actions aren't assumed and time-short recruitment teams don't overlook essential parts of the verification process. Machin adds: "To the untrained eye or someone who has just been asked to look at a photocopy of a document, there may be vulnerabilities. Some organisations and sectors, such as construction, end up with high fake rates, or employers are targeted if it's known they don't perform proper checks." Automating checks does not mean that fakes go unchecked – if there are questions over the veracity of documents or specific regulatory requirements, they can be escalated to a manual check. For the candidate it makes the experience more seamless, too. Allan adds: "Candidates find it incredibly frustrating when they are repeatedly asked to enter the same information. This increases the risk of delays in the process or losing an applicant's interest. Leading organisations are investing [in background screening technology] to achieve a seamless candidate journey with no need for candidates to enter information in some circumstances. In turn, this is driving process efficiency savings."

The five types of background checks that an employer might make

- Criminal record checks: these must be proportionate and relevant to the role in question, according to Rehabilitation of Offenders Act 1974. Employers can use the DBS to find details of any convictions – and these are a prerequisite for many roles in sectors where candidates will be working with children or vulnerable people. The UK government recently announced it would make changes to how it defines spent convictions so that certain convictions will no longer automatically appear on DBS checks
- Educational credentials: an organisation may want to check an applicant's university degree, technical qualifications, or school qualifications to ensure they meet skills requirements
- Right-to-work checks: all employers have a legal obligation to carry out checks to ensure that an applicant has a right to work in the UK, and failure to do so could result in a fine, while repeated breaches could lead to imprisonment. For British nationals this proof of right to work in the UK tends to be a copy of their passport, which can be checked in-person or digitally via accredited identify verification technology
- Credit checks: financial services organisations in particular carry out credit checks on candidates to ensure that they do not have a history of financial mismanagement, and that they do not pose an increased risk when it comes to handling money or sensitive data
- References: reference checks are a common way for employers to check that the employment history of an applicant rings true. Some companies have policies governing the information they can share in references to ensure consistency and to ensure they only share information that is factual and accurate, rather than subjective

"If there is an investigation you would need to be able to show you had processes in place and that you took the required steps to gain documentation," explains Naomi Goldshtein, a senior manager at immigration consultancy Fragomen. "Systems can range from major HR systems to bespoke systems that track immigration status and can generate reports and send expiry reminders." When it comes to someone's right to work in the UK, employers that have not performed the correct checks, or did not do them properly, could face a fine of up to £20,000 for each illegal worker, so it's crucial to keep on top of this documentation.

Right-to-work checks, in particular, are in the spotlight because, on 1 January 2021, automatic freedom of movement for EU nationals (or UK employees working in the EU) technically came to an end as part of the Brexit process. This currently means employees from EU countries will need to show proof of residence, or evidence of pre-settled or settled status. There was a six-month 'grace' period up to the end of June 2021 to enable them to apply for the status, but without it they could be working illegally from 1 July 2021. The government planned to make the status a digital document so that workers are less at risk of losing physical documentation – and this will also make it less susceptible to forgery, according to Goldshtein. Biometric resident permits (BRPs) are currently one of the most common ways to show someone's right to work, but these are still a physical piece of evidence.

While typically employers have relied on new recruits or candidates turning up to the office with physical proof of their passport or other right-to-work documents, this was relaxed during the coronavirus pandemic. Hirers were able to see the documentation over video link, together with a scan sent by email. From 1 October 2022, however, employers will no longer be able to use the temporary Covid-adjusted right-to-work guidance. This means that they won't be able to ask employees to share a scanned copy or photo of their documents, and then confirm their identity over a video call. Instead, they'll need to use an accredited Identity Service Provider (IDSP) to perform checks on holders of in-date British and Irish passports

We could see future moves towards digital identity and qualification checking in the future – with blockchain being the technology to watch. Blockchain works as a digital 'ledger' of information, shared across a network of computers with security built in at each stage, making it difficult to falsify or hack. In the hiring process, it could be used as a repository for a candidate's employment history, evidence of their qualifications and any other pertinent information. In theory, blockchain could share up-to-the-minute data about applicants' previous job titles and any outstanding grievances or courses they've completed. With everything in one secure place, this removes the need to trawl through LinkedIn and source multiple documents and references from previous employers and public bodies. While it's likely to be some years before blockchain is adopted as a secure record of employment history or qualifications, the technology is surging in acceptance – a Deloitte survey showed a 2000% increase in interest in the technology since 2013.[34]

Performing rigorous background checks on candidates is not just about avoiding a fine or reducing the risk of crime; it's also about building confidence that the candidate is giving a true version of themselves during the hiring process. Allan says that while not carrying out the appropriate checks can lead to fines or penalties for failing to adhere to regulatory requirements, the overall damage can be "much more profound and long-lasting". For example, he says, "Criminals can infiltrate a business and use it to create 'ghost employees' to siphon off small amounts of money over time and steal customer data or intellectual property. Failing to carry out adequate checks can put colleagues, clients and the general public at risk, depending on the industry." The negative publicity associated with any crime could have long-term commercial consequences, too.

34 Deloitte's 2019 Global Blockchain Survey, Deloitte, 2019, https://bit.ly/3K73GeG. Accessed 16/03/2022

Digging deeper into job responsibilities and CV claims can help assess whether a candidate will be a good addition to an organisation's culture, as well as provide an idea of their level of skills and experience. Machin concludes: "Yes, there's a legislative requirement to produce certain documents, but if their documentation is good and real it means you can have confidence in that person. It's about knowing who you're employing."

Five key takeaways

- Robust checks on potential employees help you understand more about who you're hiring and can mitigate the risk of internal crime
- Failure to carry out proper right-to-work checks could carry a fine, so make this a priority
- Having a system and audit trail in place will work in an employer's favour should there be an investigation
- Be aware of the changes to UK immigration rules that were introduced in 2021, which require organisations to ask for proof of settled status from all non-UK workers and meet a points-based threshold for skilled worker recruitment
- Fake documentation or qualifications are more common than you think, so consider investing in a third-party background screening service

Points-based immigration: what you need to know

Free movement to the UK ended on 1 January 2021, and a new points-based immigration system that applies to skilled workers from EU and non-EU countries was introduced. Under the system, skilled workers coming to the UK to work must meet a certain points threshold to qualify for a visa. Employers that hire skilled workers from outside the UK will need a sponsor licence from the Home Office; they do not, however, need to be a sponsor to recruit Irish citizens or anyone from the resident labour market with an existing right-to-work in the UK. This includes EU citizens with settled or pre-settled status, and non-EU citizens with indefinite leave to remain in the UK.

In order to use the skilled worker route, candidates must be able to demonstrate that they meet the requisite requirements, which are:

- A job offer from a licensed sponsor employer
- A job at the required skill level of RQF 3 (or A-level) and above
- That they speak English to the required standard
- A salary of £25,600 or above (£20,480 for certain shortage occupations, or if the applicant has a PhD relevant to the role)

The government has introduced several other routes to work in the UK since the main system came into force. These are:

- A graduate visa available to international students who complete a degree (or postgraduate qualification) in the UK from 2021. They can stay in the UK and work at any skill level for two years after completing their studies
- A seasonal worker visa that enables certain occupations in farming, horticulture, food processing and HGV driving to work in the UK for up to six months, providing they have an employer sponsor
- Start-up or innovator visas for entrepreneurs who plan to set up a business in the UK that is 'different from anything else on the market', and meet other eligibility requirements

- Intra-company transfer visas for existing employees who move to the UK office of a company, providing they meet certain salary thresholds and eligibility requirements, and the employer is recognised as a sponsor by the Home Office

For more information on these routes, visit www.gov.uk. All information was correct at time of publication.

If you hire groups of workers from outside the UK or plan to, keeping track of the progress of acquiring a licence, securing visas and storing candidates' settled status details is crucial. Should the Home Office wish to query any aspect of your recruitment processes, being able to show an audit trail of visa applications and confirmations will support your case.

Onboarding: you only get one chance to make a first impression

If you thought that an onboarding tool was a 'nice to have' technology investment, it might be worth bearing in mind that 4% of employees have left a new job after their very first day.[35] From the moment you make a candidate a job offer, through to their first weeks and months in a role, this experience can shape not only how they feel about your company but the likelihood they will stay and become productive in what they do.

After investing significant resources in finding, attracting and assessing that candidate, if your welcome is inconsistent or fails to live up to their expectations, it's more likely they'll feel disillusioned. More than two-thirds (69%) of employees are more likely to stay with a company for three years if they have a great onboarding experience, according to culture and engagement specialist OC Tanner.[36] And it's a virtuous circle, too, as new employees that have a great onboarding experience are likely to offer the same courtesy to new people who join after them. That 4% who didn't last the first day? There's every likelihood they'll share that experience with their friends and professional networks, leave a negative review on Glassdoor, and avoid buying your products and services again.

---

35 Bersin by Deloitte Research: Strategic Onboarding Can Help New Hires Get Off On The Right Foot, 2014, https://prn.to/3Fqx7Ve. Accessed 16/03/2022

36 An onboarding checklist for success, OC Tanner, 2018, https://bit.ly/3GpB0Ly. Accessed 16/03/2022

There are many practical advantages to offering a smooth onboarding process. Employees can gain an understanding of the organisation's structure and where their role fits in, they can meet colleagues (whether in person or digitally), and complete any relevant forms or compliance training before their start date. There are advantages in terms of speed, too. Time-consuming tasks such as providing right-to-work documentation or a P45 – that might typically be reserved for the first day – can be done online using secure document-sharing software, and reminders can be sent out automatically without the recruitment team getting involved. Onboarding specialist Sapling estimates that the average new hire has three documents to sign and 41 administrative tasks to complete on top of the 10 learning and work outcomes they're expected to achieve in their onboarding period,[37] so achieving some of this before the first day can create significant productivity gains. New employees can get straight up to speed with job-related tasks and forging relationships with colleagues and clients, rather than spending time filling out forms or completing a mandatory health and safety course.

With this in mind, more and more organisations are opting for pre-employment hubs that tie together the 'essentials' (such as contracts and policies) with learning about the organisation's culture or 'how we do things around here' and creating a warm welcome. After accepting an offer at tech giant IBM, for example, new hires gain access to a 'Soon to be Blue' online community (referring to the company's 'Big Blue' nickname) where they can learn more about its history, corporate culture and other need-to-know information. The engine powering these hubs could be your recruitment system or specialist onboarding software, but all the candidate experiences is your branding – hopefully consistent with the look and feel of their hiring experience so far. Cloud-based systems allow new recruits to access onboarding content from any device, only

37 Sample Employee Onboarding Process, SaplingHR, October 2021, https://bit.ly/3riVPBY. Accessed 16/03/2022

opening more comprehensive access to your core HR system once they're a fully-fledged employee. Lucy O'Callaghan, talent manager at Ciphr, says making processes as intuitive as possible reassures candidates that they have made the right decision. "We put new hires on our onboarding system as soon as their contract comes back," she explains. "They have access to our policies, their benefits, and their induction plan. Then on day one their user model changes so they can start booking holiday or accessing our careers hub – having navigated it before, it's prepared them for coming into the business."

It may seem like common sense, but those few weeks between making an offer and an employee's first day at work can be risky. It's often the point at which a hiring team hands over their candidate to HR or line managers, and is a common period for 'ghosting' – ceasing all communication without warning – particularly where candidates have long notice periods or have received more than one offer. As with the rest of the candidate engagement process, it can help to break down the onboarding journey into touchpoints: these are opportunities to either make your new recruit feel part of your organisation's culture or to turn them off completely. "Create a Post-It note for each touchpoint and ask: 'Is this the best we can do?'," advises McDonagh, co-founder of hireful. "Can you personalise emails, share a link to your Glassdoor reviews or to interviews with your senior people, for example?" Small 'events', such as signing an offer letter, should be as intuitive as possible – a digital signature they can complete on their phone, for example – and are all opportunities to make a good impression.

Breaking down the process into individual elements can also help you to determine where technology can make the experience more seamless, and where there is value to be had in the human touch. Technology should help to make the experience frictionless, but not to the point it becomes impersonal. So a chatbot can replace a recruiter for those 'night before' questions such as 'Where can I get my access pass?' or 'How many days' holiday do I get?' but also free up time for someone on the team to pick up the phone to a new starter who has a more pressing or personal concern.

Different stages of the process can also trigger important timesaving actions for stakeholders outside the recruitment team. For example, your HR system could send an automatic notification to the IT department that a new employee needs a laptop configured with particular software, or to order a uniform in a certain size. Onboarding journeys will differ depending on role, seniority and geography, and many tools will enable you to drag and drop features so you can curate and configure onboarding processes for different types of employee. There are also many aspects of the process that can be handed over to the new employee: self-service elements allow new starters to add and edit their own details, for example, minimising errors for HR while ensuring data is captured consistently.

Johnny Campbell from SocialTalent says that candidate relationship building must not stop once an offer has been made. "It's crucial to have regular touchpoints right up to the first day," he says. "Once the recruit has built a relationship with their manager you can hand over – the danger is dropping the ball before then." In their book *The Power of Moments,* Chip and Dan Heath use the example of the first day experience at the Asian offices of manufacturing company John Deere. Once you've accepted an offer, you receive an email from a 'John Deere friend' who introduces themselves and shares basics such as where to park, the dress code and who will be there to meet you on the first day (including a photo so you know who to look out for). Once you arrive, a flatscreen monitor in the reception welcomes you and there's a banner next to your desk saying, 'Welcome to the most important work you'll ever do'. Your first email is from the CEO, including a video about the company's mission, and there's a welcome gift of a replica John Deere plough. The scene is set to meet your colleagues, manager and their manager and the company has made that 'defining moment' of the first day at a new job a memorable one.

Of course, moments such as these have become more challenging since the coronavirus pandemic took hold, with most organisations forced to shift their onboarding experiences online. As virtual onboarding potentially becomes a feature of recruitment in the longer term, there are still ways to make the experience feel personalised despite it not being in person. "You can offer a 360-degree tour, offer invites to business briefings or town halls, or even Friday drinks over Zoom," says Nicola Sullivan, solutions director at onboarding platform company Meet & Engage. "The system can automate the output of content but you can ensure there is face-to-face engagement over video."

As should be the case for a face-to-face induction, consistency is key, she adds. "If you bring people onboard virtually this could be across different offices and countries, so you need to ensure this doesn't just come down to the quality of the hiring manager." Technology can offer those nudges to take actions that – if forgotten – might endanger the quality of that experience, from simple reminders about start dates and times to suggestions of how to make that experience memorable. The data produced by these tools can also help organisations gauge overall satisfaction with the experience, for example through pulse checks or analysis of how new employees engage with onboarding content, so HR teams can tweak and personalise the experience for future hiring programmes.

A virtual onboarding experience can also be a way to show new employees more of the culture of the organisation. Integrating recruitment and onboarding modules with an LMS (whether as part of an existing system or a third-party application) means that employees can access relevant content before their first day – whether that's a welcome video from the CEO or an interactive tool that takes them through potential career paths at their new employer. It's also a prime opportunity to engage them with any mandatory learning they need to complete, such as health and safety training. If you're onboarding a cohort of employees – for example, a new graduate intake – why not set an induction project that encourages them to collaborate while also getting to know each other and familiarise themselves with the culture of the company? How new employees engage with this content can be tracked and later linked to metrics such as retention and performance, so there is business value in offering rich learning experiences on top of your usual welcome process.

Finally, don't restrict your onboarding strategy to new hires. Every time someone starts a new role internally, receives a promotion, or returns from a period of extended leave is an opportunity to re-engage them with the business. Robert Zampetti, a partner at consulting firm EY specialising in HR transformation and workforce experience, puts it this way: "When you are onboarding someone, there are administrative transactions that have to happen over a certain period. But, in reality, our career is a series of onboardings. You do it when you change jobs, when you become a manager for the first time. It's all about the mechanics and the institutional behaviours that you experience around those changes, which either go smoothly or they don't. You either see good cultural attributes or some of the worst, and each time the organisation is put to the test."

Five key takeaways

- Think of candidates as consumers: a bad experience could put them off your organisation – and they'll tell their friends too
- Enable new hires to complete administrative tasks on your HR system as early as possible, as this means faster time to productivity
- Use onboarding to offer a window into your organisation's culture
- Make the onboarding process seamless – this can mitigate the risk of candidates 'ghosting' your company and pursuing another opportunity

- Don't limit onboarding to new hires – each move within your organisation is a chance to reinforce your culture and values

PART THREE

EMPLOYEE EXPERIENCE AND ENGAGEMENT

The evolution of employee surveys

The pandemic has been a 'pivotal moment' in employee research, according to Vijay Mistry, senior consultant at People Insight, an engagement consultancy. With employees facing unprecedented levels of change, many organisations realised the value of capturing how they felt, listening to their concerns, and acting on their responses. Organisations that had previously relied on a single, annual survey to produce an index of employee engagement began running regular 'pulse' questionnaires on topics from mental health to laptop provision. "The value of employee surveys is shifting; leaders see that they can revise questions as needs change. It's forced agility into listening strategies and we'll continue to see this grow over the next few years," he predicts.

Pre-pandemic, employee listening strategies had been evolving, but at a far slower pace. Annual surveys had long been the staple measure for organisations to benchmark employees' feelings on a range of aspects of their work life, with some offering the option to place a figure on satisfaction levels: a reduction in engagement would then prompt an investigation into why. But one of the criticisms levelled at annual surveys is that their sheer size makes it difficult to respond to the results in a timely way. "There's a big build-up and the survey itself is a lengthy process, then by the time you've been through the output, analysed the results and pushed that out, most companies have moved on," says David Godden, vice president of sales and marketing at Thymometrics. "It's a very employer-centric way of doing things."

At the same time, data analytics and communication technology has moved on to the point where it's possible to gather feedback from employees on a more regular basis without this being an onerous data-collection task. "Surveys were often an annual requirement purely because of the time they took," says Mistry. "Now they're dictated by speed. Organisations can ask certain groups particular questions, can respond quickly, and act more quickly." So while many employers still value a yearly survey as a way of 'drawing a line in the sand', or as a way of aligning employee feedback with the financial calendar or annual report, there's a growing preference to adopt a hybrid approach that combines this with smaller listening exercises that can be both more regular and focus on specific topics.

The advantage of adopting a more agile approach to surveying employees is the speed with which organisations can take action. According to Gallup, disengaged employees have 37% higher absenteeism and 18% lower productivity – costing employers the equivalent of about 34% of their salary.[38] Fewer questions focusing on specific areas, asked on a more regular basis, can give a real-time (or near enough) picture of heat spots where employees might be having trouble. Interventions can happen quickly rather than small problems being left to grow into something far more unmanageable. "Organisations with highly engaged teams achieve better business outcomes, including higher customer satisfaction, retention, productivity, and profitability – all of which are essential in executing successful growth or transformation strategies," says Patrick Cournoyer, chief evangelist at employee engagement company Peakon. The company's research has found that a one-point increase in engagement score correlates with a 4% increase in customer satisfaction. "Business leaders are eager to know when engagement dips, and understand why, and when it spikes, so they can understand what worked," he adds.

---

38 How to calculate the cost of employee disengagement, LinkedIn Learning, 2017, https://bit.ly/3GtIFbL. Accessed 17/03/2022

Some examples of focus areas could be diversity and inclusion, views on a new product release, or which benefits are the most useful for remote workers. As employees navigate the shift to hybrid working, short surveys will give managers an idea of location preferences, concerns about safety and what the organisation can do differently to sustain engagement. Peakon has recently released a tool that helps organisations measure their approach to diversity and inclusion, for example. Employees can submit their demographic data anonymously and in compliance with data protection regulations, and managers can analyse employee feedback split by population to identify any issues affecting particular groups, as well as flagging up any incidences of misconduct. "By accurately tracking meaningful metrics and acting upon issues, businesses can use this to understand the impact of their employee experience and drive better business outcomes as a result," says Cournoyer. By mapping survey metrics onto demographic data supplied by employees, or HR system data such as employee turnover and absence, organisations can see issues that need addressing or trends that could impact the business in the longer term.

Whatever form an employee survey takes, its value will be limited if managers fail to follow up on the findings. Melissa Paris, lead people scientist for EMEA at Culture Amp, argues in favour of a "continuous listening strategy" across the employee lifecycle – from asking new recruits about their onboarding experience, through to 360-degree feedback for leaders and exit surveys when someone leaves the company. The in-person conversations managers have with employees on the back of the data are even more important than the data itself. "It's about having a continuous element of improvement," she says. "Increasing the speed and scale means you can shift your culture much faster. In the hands of managers, the shift in culture can be exponential." Paris advocates placing data in context in the hands of managers so they can take action that's relevant to their own teams, and also moves the organisation towards a wider goal. "This way it doesn't feel like a behemoth task;

they feel connected because they're not the only ones dealing with something and they can see where they sit. If they try something and it doesn't work, they can iterate and move on."

Greater levels of automation make it much easier to cascade information to managers and to see whether interventions are working. "Before, everything was offline so a manager might be sent a PDF with a suggested framework for action," explains Mistry. "Now it can be seamless because you can send reports out via email, you can build dashboards and distribute tools they can use to discuss issues with their teams. We've developed online action planning tools, for example, so you can select the questions with the highest influence on engagement, log your actions, and set start and end dates for interventions." This level of visibility of how the data is being used creates a virtuous circle of buy-in from managers (who find it easier to act upon and track); employees (who see their feedback being taken on board); and leaders (who get a real-time, helicopter view of how employees feel). Many organisations reported increasing their survey activity during the pandemic as a way to 'check-in' on how employees are coping, targeting specific areas such as mental health, views on returning to the office, and practical issues around working from home.

The increased use of AI in gathering engagement data will mean even greater transparency for employees and managers. "If you make results immediately available to everyone, you can see in real-time what people are saying about the company, and this leads to a conversation," says Matt Stephens, founder of Inpulse, an engagement analysis tool that focuses on employees' emotions rather than asking them to 'rate' their experience on a scale. Participants are asked to name two emotions, a dominant and a secondary one, and to express why they feel that way. An algorithm reads the responses and places them into one of 22 'themes' which include aspects of working life such as wellbeing or the physical environment. The organisation can immediately see if employees feel strongly about something – for example, a new payroll system – and investigate further.

Similarly, Thymometrics' technology enables companies to offer employees an open (and anonymous) route for feedback at any time, rather than at set points during the year, or when managers decide. "Someone can let off steam and express their feelings anytime because it's a simple tool they can use on their mobile," says Godden. "The always-on approach means we can see trends over time."

As surveying evolves at pace, thanks largely to the pandemic, we'll see greater focus on user interfaces and smartphone-friendly apps and tools, predicts Mistry. "We'll see more apps being developed, where employees can use their phones to record results and managers can track their dashboards on their phone." But organisations should not jump into investing in engagement tools because they have an enticing consumer-like interface, he warns. "We're seeing tools emerge that use star ratings and emojis, or chatbots that ask questions as they might on a consumer website. They're a more interesting user experience but, at the end of the day, it's still just a survey. If your organisation is just getting used to more frequent listening, it's not necessarily the right time to add in new methodologies. Once you've embedded a programme and culture of listening and action, then you can introduce something different."

Of all the engagement tools available to organisations, surveys are the most established. Their form is becoming shorter, more regular and more agile, but they continue to provide a valuable benchmark for organisations to gauge the buy-in of their workforce. But whatever the mode or interface, what employees really want to see is how their managers respond to what they tell them. There is a virtuous circle where employees feel comfortable to be honest when asked, where an organisation identifies what it could do better and acts upon it, and the feedback continues. Technology on its own cannot support culture change, but it can be a useful way of giving employees a voice.

Five key takeaways

- The pandemic has heightened the need to listen to employees – think about how you will sustain this in the long term
- More regular or targeted surveys make it easier to act upon the insights
- Think about your employee listening lifecycle – are you capturing feedback at all stages?
- Empower managers by automating feedback mechanisms and developing frameworks for action
- Communicate your follow-up. Employees want to know how the organisation will act on any problem areas they have highlighted

Types of survey

Annual surveys are only one gauge of how employees feel about their working lives. As well as a growing number of employee listening tools (covered in other chapters), here are some of the most common types of surveys:

- Annual review survey: these often consist of 60–80 questions and can take employees up to half an hour to complete. While they provide a useful benchmark of how employees feel at that point in the year compared with others, the size and scope of these surveys means that analysis and, therefore, any follow-up of the results can be slow
- Company culture survey: this could be a set of questions asking employees how well they feel the organisation's behaviour matches its values. Identifying gaps can help leaders improve strategies for inclusion or change management

- Pulse survey: 'pulse' can describe a more regular feedback cycle than an annual survey, with many organisations choosing to run these surveys quarterly. They can also be ascribed to particular topics, such as how employees have responded to a new benefits platform or how they think their manager dealt with a change in circumstances. The key feature of a pulse survey is that it is short, making employees more likely to complete it
- Onboarding survey: improving the candidate experience can lead to lower employee turnover and fewer candidates dropping out of the process, so a short survey after the recruitment and induction process is complete can offer valuable insight
- Exit survey: many organisations favour an exit interview, but an anonymous survey can prompt employees to give more honest feedback
- 360-degree survey: these individual feedback surveys hold a mirror up to managers and can be used as a basis for leadership development

Sentiment analysis: from the consumer world to the workplace

During its investigation of Enron – one of the biggest corporate collapses in history – the Federal Energy Regulatory Authority gathered around 500,000 emails generated by employees. The signs had not been picked up at the time, but data analysis of the content of these emails years after the company filed for bankruptcy, in 2001, tells a salutary tale.

When data analysis company KeenCorp tested its own software using the email dataset, which showed communications between the company's top 150 executives, its algorithm assigned index scores to various points in time. The lowest occurred when the company filed for bankruptcy, which made sense. But there was also a high level of tension in 1999, 30 months before the bankruptcy filing. After further investigation, these were found to be linked to the company setting up partnership companies that would mask its losses – a transaction that eventually sparked an investigation into accounting fraud and triggered Enron's demise. What executives were prepared to say in emails clearly diverged from what they would say on the record. The former chief financial officer Andy Fastow has since said that "with the benefit of the KeenCorp Index, Enron's board of directors would have been alerted to reconsider its decision and prevent a cultural and financial meltdown."

This may be an extreme example, but this type of sentiment analysis can be a way of understanding employees' moods and perceptions and to pre-empt reputational crises, gauge if an initiative is not landing well, or even detect health and safety risks. The practice has been around for some time in consumer circles, combing through unstructured customer feedback and determining if it is positive or negative based on a score. Using text from multiple data sources such as emails, social media or review sites, the tools use natural language processing to break down what individuals are saying, placing the words into context and detecting whether they reflect positively or negatively on a brand. So a 'sick burn' might be concerning if you sell medical supplies, but a vote of confidence in the gaming community.

In the workplace, sentiment analysis could be used to identify several issues, such as:

- Clusters of disenchantment in teams, when linked to performance figures
- Reputational risk, such as rogue employees spreading rumours or sharing confidential information
- Pre-empting health and safety breaches – for example, if incidences of people stating they do not feel safe increase
- Perceptions of leaders, perhaps after a change in senior management
- Flight risks – determining if high-value employees may be about to leave
- Diversity and inclusion – for example, whether certain groups dominate particular departments and the impact this has on others
- Detecting fraud and misconduct, or compliance breaches

Data could be drawn from multiple sources, including:

- Emails
- Internal social media or networking platforms such as Slack, Workplace by Facebook, and Microsoft Teams
- Annual employee surveys and other questionnaires where there is a free-text option

- Chats on WhatsApp or other employee communication channels such as Skype and Zoom
- Exit interviews or feedback from candidates on the recruitment process
- External social media such as Twitter or websites such as Glassdoor

A key benefit of using sentiment analysis with employees is that – unlike with annual or pulse surveys – HR teams and managers can make changes based on immediate feedback and address concerns that might not have been aired explicitly. This could be in response to a new policy or a change in employee benefits, or to give a general feel for where the workplace culture is heading. "All the indicators of behaviour can be found in language, helping you to detect an employee problem before it gets worse, or the opportunity to do something about it," explains Ed Juline, head of business development at KeenCorp. "We tend to measure employee engagement in the rear-view mirror rather in real time, and predictive analytics means you can see if the signs are there." KeenCorp's technology is based on psycholinguistic research and can analyse organisations' data sources against business priorities. All the data is analysed at aggregate level so no one is personally identifiable, and machine learning ensures interpretations become more accurate over time.

One of the issues with traditional engagement surveys, even if carried out on a regular basis, is that employees often say what they think the management wants to hear, adds Juline. Sentiment analysis circumvents this by working in the background, mining multiple data streams for comments and trends, rather than asking employees outright. This was particularly important during the pandemic, according to John Sumser, principal analyst at HRExaminer.[39] "More so than ever, employees feel pressure to tell employers what they want

39 It's time for new approaches to engagement, John Sumser, Human Resources Executive, May 2020, https://bit.ly/3nnJITj. Accessed 17/03/2022

to hear," he says. "To really understand what's going on in the workforce, you need multiple data streams. Asking employees is a good start. But we need different questions, to be willing to listen to what employees have to say and to be committed to doing something to address the issues we find. Interpreting and validating survey responses requires a level of depth that is best executed with intelligent tools."

The Covid crisis has also underscored the importance of listening tools, according to Cournoyer from Peakon. An analysis across its customer base showed a significant uplift in employee comments around wellbeing as countries were forced into lockdown in March 2020. "Organisations using the platform were able to see this and move fast to address the problem. Our AI functionality meant that vast numbers of comments could be quickly and anonymously analysed to reveal the real problems. Leaders could then implement initiatives to better support their employees," he explains.

Algorithms can also be deployed to predict certain behaviours – for example, if an employee is showing signs that they could leave. Peakon's platform has an 'attrition prediction' algorithm that can analyse employee communications for keywords commonly shared before exit interviews, meaning HR can predict someone's flight risk up to nine months before they hand in their notice. Cournoyer adds: "This provides HR leads with an accurate forecast of turnover risk among different employee groups – such as departments, teams, and office locations. They can then adapt their retention and recruitment strategies before attrition becomes a bigger problem." In compliance-focused sectors such as financial services and healthcare, sentiment analysis can provide an early warning sign of risk, raising flags if certain keywords are triggered that could point to fraud or harassment, for example.

It is possible to perform some level of sentiment analysis without AI – keyword searches can be used to sort conversations into themes for review – but automation and machine learning make the process much quicker and more reliable. Even with a level of automation, organisations need to feed data into the sentiment analysis algorithm so it can 'learn' how certain words and statements are affected by context, explains Godden at Thymometrics. "It can be a challenge as companies don't always have the time for this, and reactions to things are happening in real time," he says. Thymometrics' system, which allows employees to share unstructured feedback about their workplace at any time, asks users to categorise their own comments to enhance the algorithm's understanding. This also reduces the risk of someone receiving an automated message because their comment was misinterpreted. "This brings background to the feedback rather than relying on the technology to do it. We're dealing with human beings and that personal touch is so important. People could be spilling their heart out and that warrants more than an automated message," he adds.

The benefits of being able to mine employee communications for clues about their levels of engagement are clear, but what about employees' right to privacy and potential data protection issues? A recent report by the Trades Union Congress found that just 31% of staff had been consulted by their employer around the introduction of monitoring technology, and 56% felt this was damaging the trust between workers and their employers.[40] One in seven said monitoring at work had increased during the pandemic, leading to calls from UK shadow digital minister Chi Onwurah to update official guidance for employers on electronic surveillance.

On the whole, sentiment analysis tools tend to aggregate and anonymise data so individuals cannot be identified, or will hide insights from smaller teams where a sharp change in behaviour could be linked to certain people. Your electronic communication policies or data protection policies may already include information on how the organisation will collect and use employee data, but, even so, a general rule is that more

40 Technology Managing People: The Worker Experience Report 2020, TUC, November 2020, https://bit.ly/3K6f6PL. Accessed 17/03/2022

transparency and consultation is better. For example, Microsoft launched a tool in 2020 called Productivity Score, which boasted that it could deliver insights on how employees use Teams, such as who they mention in chat functions. However, it removed the ability to see usernames after accusations these insights would border on employee surveillance. That said, acceptance that organisations may use the content of workplace communications tools to look at engagement is growing: a survey by analyst company Gartner in 2018 found that 30% of employees were comfortable with their employer monitoring their email, compared to only 10% of employees in 2015. When an employer explained the reasons for the monitoring, more than 50% of workers were comfortable with it.[41]

Similarly, employees' response to sentiment analysis will depend on what the data is to be used for. Research into employee monitoring by researchers Lynn Bartels and Cynthia Nordstrom[42] found that employees tend to assess the consequences of being monitored and 'perform accordingly'. They found that performance did not improve when employees were monitored with no explanation, when surveillance was not tied to performance measures or rewards, or when surveillance was used for vague purposes. There are also limits to how much automated tools for sentiment analysis will understand. As with employee surveys, there may also be an element of 'saying what the manager or company wants to hear' if employees know their communications are being monitored. And what about sarcasm? If the tool is powered by machine learning, the algorithm will memorise certain datasets and contexts in which certain words appear. A study of social media sentiment analysis by academics at a computational linguistics summit found that these tools could memorise if, for example, a tweet was sarcastic

41 The Future of Employee Monitor, Brian Kropp, Gartner, May 2019, https://gtnr.it/320TK5i. Accessed 17/03/2022

42 Examining Big Brother's Purpose for Using Electronic Performance Monitoring, Lynn K Bartels and Cynthia Nordstrom, Wiley, 2012, https://bit.ly/3zTtFRW. Accessed 21/03/2022

because an operator told them so, but might not recognise sarcasm again because the context would be different.[43] This also raises questions about how inclusive sentiment analysis is – if the team creating the 'rules' behind the analysis is not from a diverse range of backgrounds, the algorithm may be fixed on looking for aspects of language common to that group, but not sensitive to the nuances in other cultures, generations or those who could be considered neurodiverse.

Sammy Rubin, co-founder of health insurance and benefits platform YuLife, argues that communicating to employees that the data insights are for their benefit can increase buy-in. "Generally people feel that if they are getting rewarded, they are happy to share their data," he says. YuLife customers receive anonymised reports of how employees are using services so they can identify trends, but these reports would never show up personal mental health issues, for example. "These are a trigger for a conversation rather than a definitive answer. People downloading mindfulness apps may mean people are talking more about mental health or that work is too stressful. It just means you can investigate more," Rubin adds.

As with any engagement tool, sentiment analysis works best when used with data from other sources such as pulse surveys, focus groups, or one-to-one performance conversations. As the Enron story shows, however, its key selling point can be in spotting shifts in feeling that employees might never reveal explicitly – providing a depth of analysis that other methods might not. With organisations moving to longer-term remote and distributed working, they can identify issues and respond quickly, holding on to talented employees and minimising risks.

43 Identifying sarcasm in Twitter: a closer look, The 49th Annual Meeting of the Association for Computational Linguistics: Human Language Technologies, Proceedings of the Conference, January 2011, https://bit.ly/3nkO8u8. Accessed 21/03/2022

Five key takeaways

- Don't assume sentiment analysis is just for consumer brands – it can be a useful gauge of employee engagement, too
- Think about keywords or issues you'd like the analysis to detect, and the data sources you'd like to use
- Be as honest as possible with employees about whether their data is monitored and how this will be used
- Respond to any issues flagged up: leaving problems to fester will make them more difficult to address
- Use alongside other measures such as surveys to give a rounded picture of engagement

Network mapping: who's doing what – and where?

Imagine if you could physically track how employees communicate with colleagues and customers every day, and use that data to build better workspaces, improve people's experiences at work, and make sure they're working together in the most effective way. For many, this will sound like something out of a dystopian movie, but this technology is becoming more established. Humanyze, which develops wearable ID badges to map employees' interactions, has used this way of tracking employees at a European bank.[44] Every employee across a few branch locations wears a tracking badge; all have the same training and demographics. Yet looking at the data produced by the trackers and comparing it to sales data, the highest-performing branch had the most interconnected social network. The data also shows how, in another branch where performance was not as good, one tight-knit group had excluded a group of new hires. In a third branch, the fact two teams were on separate floors meant vital communication was not happening as often as it could, which had a negative impact on sales.

For the bank, this information meant they could make relatively simple, low-cost changes that ultimately saw sales go up. And while physically arming employees with devices that track their interactions is still in the early adopter phase, being able to map spheres of influence and talent digitally is

44 A European Bank Improves Performance Gaps Between Branches by 11% to $1 Billion Through Better Workplace Design , Humanyze, August 2021, https://bit.ly/3qoVBKz. Accessed 21/03/2022

well within most organisations' reach. Network mapping is gaining in importance as the structure of organisations changes from traditional, linear hierarchies to something more like a 'spider's web' of connections based on knowledge and experience rather than job titles. "Networks are replacing the old ways of working, both within our organisations and with other partners or stakeholder groups outside the company. This change in our working world is being facilitated by multiple converging outside trends: demographic, technological, fiscal austerity and legislative changes," explains Penny Scott, a network engagement practitioner and academic.[45] Whereas we used to rely on 'org charts' to show (generally linear) connections between senior executives, managers and their reports, effective network mapping can show people working across organisational boundaries, areas of skill and influence, and how these interactions affect the success of the business.

Lizzie Benton, an organisational culture consultant, says company structures are now becoming more like 'bubbles' than defined lines between top executives and those beneath them. "In mapping this more collective culture, we look at what people's roles are, what projects they're working on, and the impact of what they're doing," she explains. "It becomes a living document that people can look at to see who is working on what, or who is the best person to answer a question, rather than a job description or org chart that sits in a drawer." Many organisations use project management tools such as Trello and Asana to map and update these connections, she adds, while there is a tool called GlassFrog targeted specifically at companies with looser management structures. Many HR systems such as Ciphr HR offer organisation chart functionality that enables you to see a 'live' graphical representation of influence in the business so you can identify gaps or areas that require attention. These tools can help leaders track those who lie outside the

45 Network Mappings as a Tool for Uncovering Hidden Organisational Talent and Leadership, Penny Scott, Queen's University Industrial Relations Centre, Canada, January 2016, https://bit.ly/3foouA2. Accessed 21/03/2022

boundaries of the organisation, too, to map skills for freelancers and contractors who come in and out of the business on a project-by-project basis. If the tool is dynamic and quick to update, organisations can add these contingent workers to the network to show where their skills add value.

Consulting firm Deloitte describes this as 'organisational network analysis' or "a structured way to visualise how communications, information, and decisions flow through an organisation. Organisational networks consist of nodes and ties, the foundation for understanding how information in your organisation is flowing, can flow, and should flow." Informally, we might have 'go-to' people in the workplace who always know what's going on, and it's these spontaneous but important connections that network mapping aims to capture. For example, working out who the central 'nodes' or influencers are can help embed change more quickly.

This 'snapshot' can then be used in several ways:

- Succession planning: if this person or these people leave the business, what skills do we need to build or how do we transfer knowledge?
- Scenario planning: useful to understand how mergers and acquisitions might impact the distribution of roles
- Diversity and inclusion: network mapping can show if certain teams lack diversity and suggest what might happen if demography changes
- Assess the strength of certain relationships across all levels (and can include external stakeholders too)
- Allow leaders to see who is well connected and who is isolated, as well as identify 'hidden leaders'
- Identify blockages in internal communication
- Provide a useful roadmap when implementing new processes or policies

If traditional career progression is less linear, what does this mean for talent management? Network mapping can enable organisations to overlay data on employees' skills and ambitions on their spheres of influence with others, argues Ian Lee-Emery, CEO of Head Light, a talent management software company. "It's how we make sure someone is doing the right thing, they're on the right project, and they're developing their skills," he says. "We can use software to underpin conversations around performance and as a tool for career conversations. What are the most important interventions based on what people think is important? This is how we understand what drives and enables employee engagement." It can also help organisations build better oversight of costs or form a business case for areas of investment, says Steve Black, co-founder and chief strategy officer of Topia, a talent mobility platform. "It can help organisations see where people are in a talent matrix, how much it might cost if they gained international experience, for example, or whether they're suited to remote or office-based working," he says. Many organisations already use talent matrix tools such as nine-box grids to map skills; this just adds another dimension.

As the example of the tracking badges in the bank shows, employees' social connections can be influential in how productive they are, and their ability to meet targets or innovate. In his book *Social Physics*, Alex Pentland argues that if we can see what someone else is doing, we often follow their strategies and learn from them. On an organisational level, this pattern of learning forms a process of 'community vetting' that results in people adopting new behaviours. This can be both positive and negative: it can become an echo chamber where the flow of ideas leads to narrow ways of thinking or even errors, or conversely help 'seed' helpful behaviours through the organisation. Benton adds that as organisational structures and boundaries become fuzzier, this 'collective' influence can be difficult to grasp for some. "We're brought up on systems of hierarchy, and people are often conditioned to being in a 'position' of some kind – being 'sales director' is part of who you are. It can be hard for people to understand they no longer have dominance over others," she says. Mapping skills, connections and influence can show the organisation as a bigger ecosystem, argues Benton, where if someone adds value it also enhances value for their colleagues.

It's impossible to consider the future of networks within organisations without looking at the impact of the Covid-19 pandemic. With major companies such as Unilever announcing they are "never going back to five days a week in the office" (according to chief executive Alan Jope), what does this mean for the impact of employees' interactions and how we track them? Michal Izak, a reader in management at Roehampton University, believes the perception that remote working means greater levels of autonomy is not always true. "We may think people are more empowered as they're further away, but that's not always the case," he says. "Some managers are enforcing more direct control by asking employees to be on video all the time or sending empty emails they have to open; others leave more space to self-manage but people work harder because they're keen to be visible." Physical distance can make it more difficult to chart dynamics between teams and perceptions of power, as there are so many more variables. He adds: "There are different types of remote worker – from someone who has worked from home before and is more tech-savvy, to someone combining work with family commitments. All this influences how productive and empowered that employee is. Any tool you use to track this needs to be sensitive enough to take a number of variables into account."

In the medium term, being able to see all these variables will help leaders to plan any potential return to the office, even if it is on a more hybrid basis. Longer-term, could we see greater adoption of 'smart badges' or other trackers? One of the barriers to technologies such as this taking off is the concern many organisations have about employees feeling watched. Systems such as ActivTrak or Sapience already offer dashboards where managers can track employees' online interactions, while at the other extreme office outfitter Herman Miller now produces smart furniture that collects data on employees' physical working habits – tracking who talks to whom using a wearable device is arguably on the same spectrum. Humanyze emphasises that its data is collected on an aggregated basis rather than individual-based insights, but it will be up to leaders to be transparent about how, and more importantly why, they're gathering this information. Even on a basic level, however, using network mapping to identify hotspots of talent and influence, identify where employees need extra help, and find any blockages in communication, could be the 'glue' they need to improve life at work in a post-pandemic world.

Case study: How HomeServe maps skills

Talent planning is a strategic priority at home repair company HomeServe. Because it is undergoing a period of significant growth, the company needs to be able to see where skills are geographically and where it needs to attract and develop talent.

Traditionally, HomeServe had relied on localised talent planning, with different countries operating different processes and recording talent and career information in different ways. "For some it was spreadsheet-based; others made use of presentation slides or [Microsoft] Word-based formats. There was no single process – and no single repository which stored all the information about our key talent," says Chris Fenton, group director of talent and development.

The company brought in the Talent Successor tool from Head Light – a central online system that HR and managers can access to update information about 200 people identified as key talent. Their profiles include demographic and biographical information as well as details of their career aspirations and options for mobility. The system plots individuals into a nine-box grid based on their performance and potential ratings – something that had previously been done manually with each employee plotted separately. It can also produce a global talent report so HomeServe can see where current talent is, their readiness for a career move, and identify areas ripe for development, meaning everything it needs for succession planning is in one place.

Five key takeaways

- Organisational structures are evolving to become more network-like: network mapping can help you see who makes decisions and how
- Check if your HR system offers organisational chart mapping, or find a project management tool that can plot connections and statuses
- Track informal communications or relationships (either in the office or via platforms such as Slack or Teams) – do these social networks have an impact on performance?
- Use network mapping to plan how different teams interact, taking into account if they will be working remotely
- Keep an eye on future developments: physical connection trackers or 'smart badges' are beginning to gain ground as a way of tracking productivity

Wellbeing at work: what works best?

When Moodbeam launched a silicone wristband to tell managers if their workers were happy or sad in early 2021, it provoked a mixed reaction. The wearable technology consists of two buttons; an employee presses yellow if they are happy, and blue if not. Throughout the week, the wristband sends data to an online dashboard so managers can track the wellbeing of their teams. The company claimed this would create a vital connection between colleagues at a time when most were working remotely. Critics described the wristband as a gimmick that employees would refuse to use and an intrusion into their personal lives. Whatever your stance on its usefulness, it's part of a rapidly growing market of wellbeing apps, wearables and platforms that organisations can choose from to support their staff.

The global workplace wellbeing market is growing at a rate of around 7% a year, according to consulting firm Deloitte, and is expected to reach a value of $90.7 billion (£66.4 billion) by 2026.[46] It covers a vast range of interventions, from workplace benefit platforms and fitness subscriptions to innovative wearables that can track employees' movements and even their sleep. There are solid commercial reasons to prioritise wellbeing: Deloitte figures show a return on investment of £5 for every £1 invested in mental health, for example.[47] Many employees already use

46 Designing work for well-being, Deloitte, May 2020, https://bit.ly/3zUmTvn. Accessed 21/03/2022

47 Mental health and employers: refreshing the case for investment, Deloitte, January 2020, https://bit.ly/3zWOpZd. Accessed 21/03/2022

smartphone health apps or own smart watches, making them an easy sell-in and reducing the need for training. Over the course of the Covid-19 pandemic, offering a broad range of wellbeing benefits has become a way for employers to be proactive about supporting employees and showing they are valued. Almost two-thirds have increased support for staff across mental, financial or physical wellbeing during this time, according to GRiD, the industry body for group risk protection.[48]

"Pre-pandemic, employers were already making strides in this market as the world was changing fast and people were struggling to adapt to that change," says Mike Blake, director of health and benefits at consultancy Willis Towers Watson. "The pandemic has accelerated that. Wellbeing support has become part of what's expected from your employer, but the tricky thing for organisations is sorting out what to implement when there's so much available." There has also been a subtle shift in the perception of wellbeing 'ownership', believes Jim Woods, co-founder and CEO of BetterSpace, a digital marketplace for wellbeing products and services. Where once potential recruits might look for a subsidised gym membership, they increasingly seek out support on everything from mindfulness to giving up smoking. "Historically it's been regarded as 'your' issue: we pay your salary and you look after your own wellbeing," he says. "But over the last decade we've started to see employers engage more with it, driven by the need to increase productivity and to attract and retain talent." The impact of the Covid-19 crisis has also influenced employers' budgets and investments, with many organisations pivoting from in-person services such as gym classes to virtual programmes and tools. Debra Clark, head of specialist consulting at Towergate Health & Protection, explains: "Before, they might have offered a selection of benefits but didn't delve into the detail as much. The problem is how do you get to everybody, not just those with an interest in wellbeing

48 Six in 10 employers have increased wellbeing support to staff in light of the pandemic, GriD, February 2021, https://bit.ly/3GuYFKE. Accessed 21/03/2022

already? With a multigenerational workforce, you need to think about communicating with people in different ways."

Data can make all the difference between a generic but underutilised menu of wellbeing options and something more personalised and impactful. Every wellbeing touchpoint in an organisation – whether it's how many mindfulness classes an employee takes or how often they use a virtual GP service – offers employers the chance to analyse data and spot trends. This means they can tweak the benefits or tools they already offer, build a business case for investment in more, or identify potential issues around burnout or stress before they become unmanageable. (According to analyst company Gartner, 22% of companies worldwide track employee movement data, for example.) "At an aggregated level, an employer can get a deep insight of the real wellbeing needs of staff," says Martin Blinder, CEO and founder of Tictrac, a wellbeing engagement company. "You can see what percentage are engaging with different wellbeing tools and content, in what locations. This is powerful because employers only have a finite budget but can begin to understand staff's wellbeing needs much better."

Better personalisation means higher engagement, adds Woods. "Before, [employers] might choose a few things that would appeal to as many people as possible but then only see 2% to 5% uptake for some things. Moving to a personalised or on-demand model where apps or services are tailored to employees personally can drive engagement levels as high as 80% or 90%," he says. Law firm Linklaters has a 94% engagement rate with the tools its employees access through BetterSpace, and claims to have seen a 70% improvement in employee mental health as a result. However, as the reaction to the Moodbeam wristbands showed, there's a fine line between collecting data to get an idea of wellbeing trends and intruding on employees' personal lives. In 2016, health insurance giant Aetna ran a study where it rewarded employees with $25 for every 20 days they reported more than seven hours' sleep, asking staff to sync their Fitbit or other health tracker to the company's wellbeing platform.[49] The company rebutted claims that it was intruding into their personal lives because employees self-reported the data and opted in. In fact, more than 49,000 employees took part. Most platforms collect data at aggregate level so individual employees cannot be identified, but asking employees to opt in to data collection is good practice, as is transparency over who will process the data and for what reason. One of the key drivers is what HR or managers plan to do with the data: if they identify that people in certain roles are showing signs of burnout, what measures are put in place to stop that happening, or do they offer support? Katharine Moxham of GRiD says follow-up is crucial. "It's great to see that employers are stepping up to the plate: not only do the majority understand that they have a great responsibility for the wellbeing of staff, but many are also implementing practical changes to make a tangible difference. However, we urge businesses who have either not made any changes or who have decreased support to take stock. Employees have long memories and their loyalty can be quickly won or lost during times of adversity so all employers should be playing their part in supporting staff wellbeing."

The Aetna example touches on one of the key drivers for engagement with wellbeing tools and services: the lure of an incentive or reward. While in this case it was financial (the company also paid out when employees recorded a certain amount of exercise), it's important to highlight other incentives, says Blinder. "You don't necessarily want to pay people to get healthy, but because they enjoy the experience," he says. "You could offer the chance to win something, a discount on something or a free coffee." YuLife, a new health benefits platform, gives members currency, known as 'YuCoin', for every mile they walk or cycle, or how many minutes of mindfulness videos they watch. The system links to employees' wearables such

49 The company that pays its staff to sleep, BBC News, June 2016, https://bbc.in/3qnEEQp. Accessed 21/03/2022

as Fitbits and Apple watches and YuCoin can be exchanged for vouchers for stores such as Amazon. But tangible benefits aside, YuLife also provides leaderboards so employees can compete against their colleagues, and there's a 'universe' where they can make their own avatar and complete various challenges.

"Gamification creates more social cohesion and interaction, counteracting the fact there's so little engagement with many wellbeing programmes," says Sammy Rubin, YuLife's co-founder and CEO. The thinking behind YuLife's platform draws heavily on the work of Stanford University behavioural scientist BJ Fogg[50], who argues that two of the key drivers of behaviour change are sensation and belonging. "You'll have people who just want to get the miles in and achieve, then others who are motivated by doing things with other people. Creating an emotional response can trigger someone into action," adds Rubin. Organisations can amplify this sense of 'nudge' or community by linking wellbeing programmes to social or collaboration platforms such as Yammer and Slack, particularly now many employees are working remotely, adds Clark from Towergate. "Previously, you might have shared with people in the office that you used a physio and made a recommendation, but now this needs to happen virtually," she explains. "If someone shares a success story with others, colleagues are more likely to share their own next time or feel motivated because they've helped someone. This creates momentum and keeps engagement levels going."

In the future, aggregating data across not just single organisations but beyond to whole sectors and industries could help healthcare companies pre-empt issues such as burnout or tweak the support on offer to reflect a wider need. "This market will be quantum times more efficient in the next 10 years than it is now," says Woods from BetterSpace. Being able to offer a comprehensive and personalised wellbeing package will become a point of difference

---

50 *Tiny Habits: The Small Changes That Change Everything*, BJ Fogg, December 2019

when it comes to attracting talent, too. "Millennials and generation Z employees really care about employers that can provide holistic support – they want to work there, stay with them longer, and be more productive while they're there."

Five key takeaways

- Personalise your wellbeing offers through data: what tools and services are used, and how could they be improved?
- Start small with a limited number of tools and build evidence for their efficacy: the choice of programmes and apps can be bewildering
- Consider building your own wellbeing tracker, or incorporate questions on wellbeing into other survey tools, as this can be cost-efficient
- Be mindful of how data on employee wellbeing is used: process it at an aggregate level so it is not intrusive
- Make it social: a sense of belonging increases engagement with wellbeing tools and services

Build a bespoke wellbeing tracker

While many wellbeing platforms offer data analysis and dashboards so organisations can track different aspects of staff health, another option is to build a bespoke wellbeing tracker. This could take a similar form to an engagement survey, asking users questions about aspects of the wellbeing provision they use and the value they bring.

Building a tracker for your own organisation means you can ask tailored questions about specific aspects of employees' lives such as mental health or financial wellbeing. It also means you can create baselines specific to aspects of wellbeing that are relevant to your organisation. Questions could cover number of areas, such as:

- Which wellbeing services have you used in the last three months?
- How does [name of service] improve your mental wellbeing?
- How does [name of service] improve your physical wellbeing?
- How satisfied do you feel with your mental/financial/physical wellbeing at the moment?
- How would you rate the support given to you on [aspect of wellbeing] by your manager or employer?
- What wellbeing services would you like to see added to what the company offers?

By adding a numeric scale (1 to 10) or asking employees to choose from a Likert scale (the term used to describe survey questions with a scale of options such as 'highly satisfied' to 'very unhappy'), it's possible to track their levels of satisfaction over time and identify trends in particular teams. Furthermore, there are several free or low-cost options to distribute questions such as SurveyMonkey, Google Forms and Microsoft's Forms tool (which is likely to be part of your Office 365 suite), so it's a viable option for those on a budget. Alternatively, you could use your engagement survey tool to ask specific questions around physical, mental or financial wellbeing.

From intranets to employee experience platforms: the evolution of collaboration tools

The death of office email has been declared time and time again – predictions have often been made that, with the rise of chat and communications tools such as Microsoft Teams and Slack, inboxes would be bare and employees would suddenly have much more time on their hands. Slack even sold itself as the 'email killer' when it was first launched. But while this is not a reality yet (the total number of business emails sent and received each day is expected to be more than 333 billion by 2022, according to the Radicati Group[51]), the more immediate nature and easy user interfaces of workplace chat channels or collaboration platforms has made them an increasingly popular way to communicate. This was heightened during the coronavirus pandemic, with colleagues resorting to chat channels on company messaging platforms to replace the conversations that used to happen around the water cooler. In fact, Slack estimates that between February and March 2020, the average number of messages per user increased from 197 to 252.[52]

This growth has been fuelled by a shift in employees' expectations of how they will communicate at work. Gartner's 2019 Digital Worker Survey found that 58% of respondents used

51 Email Statistics Report 2018-2022, Radicati Group, 2018, https://bit.ly/3zXpbd3. Accessed 21/03/22

52 How London has been Slacking in the lockdown, Amelia Heathman, Evening Standard, 26 June 2020, https://bit.ly/33EbELG. Accessed 21/03/2022

a real-time mobile messaging tool daily, and 45% a social media network[53] – so when they come to work, they feel frustrated if this consumer-grade experience is not available. Shimrit Janes, director of knowledge at the Digital Workplace Group, explains why this is important. "The amount of time it takes to move between user interfaces impacts our cognitive ability to get things done because we're constantly shifting context. Users crave an experience that is less fragmented and interrupted." When employees are empowered to get on with their tasks for the day, rather than faced with spending time checking email trails or moving between tools, this in turn creates a 'sense of satisfaction' that impacts positively on their engagement.

For years, employee intranets have been one of the main routes for organisations to communicate with workers and enable access to HR and benefits systems so they can book holidays or report absence. However, Josh Bersin, global HR analyst and dean of the Josh Bersin Academy, argues that the push for simplicity from employees had led to the emergence of 'employee experience platforms' that offer an entry point to other workplace systems such as payroll with a more consumer-like experience at the front end. "Consider the consumer market. Google, Facebook, and Amazon have dozens of back-end systems, yet we see a single easy-to-use interface as consumers," he explains. "Just as they have abstracted away complexity with a front-end layer, so must we build a similar architecture for employees in our companies." During the course of a day an employee might need to access a learning tool, provide some performance feedback for a colleague, and book some time off – an employee experience platform aims to bring these disparate actions together behind a seamless gateway. In February 2021, Microsoft launched Viva, an employee experience platform that brings together access to company resources such as policies and benefits, provides data insights into how employees use the

53 Gartner says worldwide social software and collaboration revenue to nearly double by 2023, Gartner, September 2019, https://gtnr.it/33wkVVS. Accessed 21/03/2022

platform, and enables access to learning materials and a way to connect with information and experts relevant to their job. (We look at this evolution in the section below.)

The need to access information or communicate 'in the flow of work' is precisely why employees turn to the more consumer-like interfaces of Teams, Slack, Google Workspace or Workplace from Facebook, says Amy Cook, a partner at workplace collaboration consultancy Cook & Co. "Users feel more autonomy if they can engage with someone or authorise something in the flow of work," she explains. "You might be working offsite and want to file an expense form on your phone rather than waiting to have access to a certain system in the office, for instance." Slack, for example, offers open APIs that make it relatively easy for organisations to connect other systems into its workflow, meaning they don't need to switch programmes to complete simple tasks, she adds.

"If you look at the tools, each of them does something really well," adds Kenzo Fong, CEO and co-founder of Rock, a platform that aims to bring workplace tools together in one place. "So you might have Slack for messaging, and Asana or Trello for project management. The problem is that all these apps are disconnected so you might have to share a link to a Google Doc or a task from your project board." With distributed workforces increasingly becoming the norm, being able to create a space where these tools come together is crucial, he argues – it might not be possible to ask the manager to 'ping' over a document as we did before, so centralised access helps to create a more frictionless experience.

The advantage of easy integration is that collaboration tools can be more easily personalised for employees: so if you need to access the team calendar several times a day, it makes sense to connect to that; if you're a salesperson who needs to access the customer relationship management system you can do this through a click of a button. However, with many organisations shifting corporate software investments into the cloud thanks to lower costs and easier systems management, some IT departments may be reluctant to let this happen. Janes adds: "With cloud-based platforms, vendors have control over the roadmap and updates are pushed out to clients, which means organisations sometimes try to limit the amount of customisation. But there are still things they can do to build a good 'user experience layer', for example by building in microservices via APIs."

These platforms need not be limited to improving productivity – they can also be a useful tool for boosting engagement. Communications about achievements or acknowledgements of a colleague's work can be amplified through these channels. Ray Pendleton, founder of Thirsty Horses, which has developed social recognition and engagement platforms for several NHS Trusts, calls this "crowdsourcing the strategy." He adds: "The strategic goals are created at the top level, but as employees interact with each other, giving each other feedback or giving someone a good rating for a job, the rest of the organisation sees their contribution to that strategy in action." So a team might be working on a project in one channel, but in that they can also see their performance objectives, relevant learning and actions that have already been taken. This opens up engagement because there is more transparency over their role in wider organisational goals, Pendleton explains. "Rather than being told to 'make 20 widgets', employees are nudged through a process of support from their managers and peers. They feel empowered because they know what they need to do to meet their objectives." Peer-to-peer feedback reinforces the message and enhances engagement.

CASE STUDY: Workplace from Facebook creates connection at BT

With an employee base of 105,000, you'd imagine that communicating with the workforce in a timely and interactive way at BT would be a challenge. However, in 2019 the telecoms company opted to roll out Workplace from Facebook as a way to communicate across its different brands and locations. Colleagues from retail shops and contact centres can use the platform to connect with engineers and support teams, for example.

The platform is used in a multitude of ways. Employees can tag teams in posts asking for help on specific topics, there are knowledge-sharing groups, and leaders can see instant feedback on announcements or initiatives through comments and reactions. Senior leaders, including CEO Philip Jansen, have used Workplace to host live interviews and Q&A sessions, and employees can give their input using its poll facility.

There are now 83,000 registered accounts across the workforce, and this is slowly increasing each month, with 54 countries on board. One of the key reasons BT chose Workplace was the familiarity of the Facebook user interface, which keeps monthly active user figures high. Video is one of the most popular elements: engineers use it to upload footage of problems they're working on to ask colleagues for advice or to share good practice, for example.

It has also played a crucial role in maintaining wellbeing and engagement during the pandemic. Colleagues have used a hashtag #TheBigStretch to showcase wellbeing ideas and a live video tool means they can have two-way conversations, which help them feel more connected. In BT's colleague survey, access to wellbeing information and support scored more than 80%, with Workplace contributing significantly to that rating.

Furthermore, the data thrown up by people's likes, preferences and shares can help organisations personalise the experience even further. Combining this with generic HR data such as a person's role, which part of the company they work for, and in which location, can ensure they see the tools most useful to them. But this data collection and monitoring needs to be kept at an aggregate and anonymous level. Microsoft faced a backlash in 2020 when its data analytics tool, Productivity Score, was shown to throw up insights such as how long employees spent in Teams meetings, whether they had their camera on, and how many times they mentioned someone in a chat channel or over email. After accusations that this crossed a line in terms of employee surveillance, the company said it would remove the ability to see people's usernames in the data. Martin Fleming, a fellow of the Productivity Institute at the Alliance Manchester Business School, is hopeful that – on an industry level – this 'big data' can be used to achieve better insights on the balance between engagement and burnout, and how technology can support this. "The likes of Microsoft and Google know what apps are being used at what time of day, and the telecoms companies know where we are at any time," he says. "There are enormous amounts of data being accumulated on workers' practices and habits over this period, and I would expect that, over time, it will be used to create some solutions."

And herein lies the challenge: how do organisations ensure a seamless collaboration experience will have a positive impact on productivity without constant notifications to employees who already feel overwhelmed? The shift to digital working during the pandemic led to many workers feeling compelled to be always available, to appear on multiple video calls or respond to messages straight away. Cook advises this is something managers need to be mindful of when rolling out collaboration tools. "If a user wants to be accessible at different times of day, we need to allow that. Just because these tools can offer instant communication, doesn't mean it should be an expectation," she says. "Make people aware (and show them) that they can snooze notifications or block time out of their calendars." Fong predicts that post-pandemic ways of working will evolve so that workers choose both when and how they communicate. "In the future, I think we'll see the most complex work – such as a recruitment interview or collaborating on a project – happening over live channels such as video, but everything else will become task-based and asynchronous," he says.

This is where the human side of engagement must come to the fore, Fleming argues. "Over the course of day and night there is an expectation that employees will respond, even where senior leaders don't set that expectation. The tools themselves create an expectation and desire among workers, even in the most understanding of organisations," he says. So while increasingly sophisticated tools emerge to help employees achieve more and collaborate faster, it will become more important than ever for HR and managers to set realistic goals and prevent burnout from driving down engagement that could push their most talented and skllful people to leave.

Five key takeaways

- Identify what different teams need to do and how they communicate – this will influence the type of collaboration tool or platform that suits
- Think about what works with synchronous communication (live chat or video meetings) and what can be managed asynchronously
- Prioritise and personalise: make it easy for employees to access the tools or messaging channels they use most
- Manage expectations: messaging technologies can create an obligation to respond immediately. Encourage employees to set boundaries
- Use data anonymously to identify which tools are most useful for different functions in the organisation

How this market has evolved

According to analyst group Gartner, the market for collaboration tools is highly fragmented and 'contextually focused', meaning users will use the tool most useful for the activity they are working on. However, with the number of knowledge workers expected to increase to 1.14 billion by 2023,[54] software providers are keen to offer as much integration with their own systems as possible. On the one hand, there has been significant growth in communications and messaging tools such as Slack and Teams. But there's also a nascent market for 'employee experience platforms' that have an element of messaging but that are similar to intranets in that they offer a gateway to tools an employee might find useful during the course of their day. Some have grown out of productivity tools such as Microsoft, while others are linked to central HR systems but boast a more seamless and consumer-like user interface.

Here are some of the main players in both camps:

Communication and collaboration

Microsoft Teams

Many organisations opt for Teams because it comes as part of the Office 365 subscription, meaning employees can sign in easily and it is linked to other Microsoft apps such as Outlook, Word and Excel. According to Microsoft, it's the fastest growing app in the company's history, with more than 330,000 companies using it. Users claim its key benefit is its integration – there's no need to switch apps if a team is collaborating on a document and it's easy to share a file during a video chat, for example. Users can create channels where they can hold meetings, add files, and link to other apps or bots via APIs. In many organisations, Teams has overtaken use of other Microsoft platforms such as Skype for Business and Yammer.

54 Gartner says worldwide social software and collaboration revenue to double by 2023, Gartner, September 2019 https://gtnr.it/33wkVVS. Accessed 21/03/2022

Slack
Bought by customer relationship management software company Salesforce in 2020, Slack is a cloud-based instant messaging tool that aims to replace email as the chief way teams interact on projects. One of its key benefits is its openness to integration with other third-party services – so employees can switch between looking at a document a colleague sent over sharing a post in their Twitter feed without switching applications or logins. Its free tier has no limitations to the number of users that can be added to a group, which is useful for businesses with smaller budgets and explains its rising popularity with community action groups.

Yammer
Microsoft has owned Yammer since 2012, but the platform faces stiff competition from Teams and other collaboration tools. It is more of a social networking platform than Slack and Teams, with many organisations using it as part of their company intranet. Users can like and comment on posts in a similar way to how they might on consumer apps such as Facebook, and it ensures that intranet content is dynamic and can be shared more efficiently. Like Teams, there is integration with other Microsoft apps such as Word and Excel so it's easy to share files and collaborate.

Workplace from Facebook
Facebook launched Workplace in 2016 with the aim of bringing its hugely popular social network to the world of business. This makes one of its key advantages its familiarity with users – most will know their way around the application so little-to-no training is required. There is an instant messaging tool called Workplace Chat that is similar to Messenger, and groups can be created to work on specific projects. There is also a multicompany feature that means staff can communicate with external

teams. Workplace offers a live video stream function so organisations can communicate with large numbers of employees at once.

Google Workspace

Formerly known as G-Suite, Google Workspace offers a similar range of collaboration apps and tools to its competitors, including video meetings, chat and the ability to collaborate on documents and forms. The platform gives employees access to a shared Google Drive for their team. In 2020, Google's parent company Alphabet said it would add more applications into the enterprise version of Google so users would need to switch apps less frequently. Its video calling software, Meet, is now accessible via email for business users. Gmail still remains more popular as a personal email option, however, with more than 1.5 billion active personal users compared to around 6 million business customers.

Employee experience platforms

Microsoft Viva

With a foothold in the office tools market and many organisations already using Teams, it makes sense that Microsoft launches a tool that integrates these elements of the employee experience together. Viva comprises four main elements: Connections (links to company resources such as policies and benefits); Insights (data for managers and employees to get insights on how they use tools to collaborate); Learning (access to learning content); and Topics (connecting employees with the people and information they need within the organisation).

ServiceNow

ServiceNow is a cloud-based platform that integrates and automates workflows and business processes in IT, customer service and HR. It offers 'employee experience packs' of preconfigured data based on common use cases (such as logging absence) and an 'Employee Service Centre' – a unified portal where employees can access information and services.

HR management and enterprise resource planning systems

Many organisations will have already invested in central HR software that acts as a system of record for employees' personal details, absence, payroll and the like. Using APIs, developers can build a link between a more consumer-like interface and these systems, creating a more integrated experience for employees. This is not an 'off the shelf' option but means it can be customised to individual organisations.

Consumerised interfaces: an Alexa for the office?

Voice assistants now occupy multiple rooms in our homes, doing everything from telling jokes to sharing weather forecasts and adding items to our shopping lists. For consumers used to the likes of Alexa and Google Home, it's become second nature to ask them to turn on the lights or set an alarm. As our work and home lives have become ever more blurred, how has this affected our expectations of the technology that helps us do our work? According to Janes at Digital Workplace Group, the chasm between the consumer interfaces we have become used to in our personal lives and the technology we use at work has created a 'user experience debt'. "There's a disparity between the private experience we have as consumers and what happens at work, leaving us frustrated and asking 'Why can't it just work like that?'," she says. "People feel that things should just work and be intuitive."

Statistics from Dell back this up: 35% of employees surveyed for Dell's Future Workforce Study[55] said that the technology they have at home is far more advanced than what they have at work. Remote and millennial employees would be more likely than other workers to quit a job due to substandard tech, and seven in ten felt it influenced their new job choices. Consumer products and services are increasingly developed and improved through design thinking – an approach where the user is the

55 Future Workforce Study, Dell/Intel, 2016, https://dell.to/3nqNxHc. Accessed 21/03/2022

focus – but much of the tech we use at work is yet to catch up. "Today I believe the biggest change is not just the role of technology, it's the need to focus on experience design for all aspects of work," says Bersin. "We can no longer think of HR as a function solely for creating and administering people-related programmes. We have to design and deliver experiences, just like we do for customers."

According to Deloitte, the key to meeting these expectations is putting the user at the heart of how we build tools and services at work. "In order to create an enduring relationship, be social in nature, and create meaning, experience must come from and be focused on the individual," the company advised in its 2019 Human Capital Trends report.[56] "When experience comes from the individual (bottom-up), it is designed starting with the employee's pre-existing tendencies to enable them to do their best work in the way that works for them. When experience is focused on the individual (personal), it is designed to incorporate all of the psychological needs that must be met in order for someone to perform their work well." That said, investment in consumer-facing chatbots far outstrips their use for employee tasks such as HR self-service and helpdesk queries.

While we're some way off the widespread adoption of voice assistants at work, many organisations are beginning to identify processes or tasks in their organisations that can be automated to create a smoother experience. These range in sophistication from simple chatbots that can route requests to back-end systems such as HR and payroll, to complex AI tools such as IBM's Watson that can analyse complex datasets, offer intelligent suggestions, and even make low-risk decisions. They are well suited to a range of HR processes across the employee lifecycle because these tasks tend to be transactional – setting up an interview for a candidate, for example, or sending out a benefits statement. The chief benefits are in the time saved:

56 From employee experienced to human experience: Putting meaning back into work, Deloitte, April 2019,
https://bit.ly/33aPuRj. Accessed 21/03/2022

technology services company Wipro launched an AI tool called Holmes to deal with IT helpdesk tickets using natural language processing in 2015. Over the space of two years, just 10% to 12% of the 5,000 daily service requests were being passed to a human and 20% of the requests were resolved end-to-end without any human involvement. Thanks to machine learning, AI-powered bots get better at handling tasks with each interaction, so time and cost savings improve and the 'human' team can focus on more strategic work.

Here are some other potential use cases for consumer interfaces:

- An employee can ask a voice assistant how many days' leave they have remaining
- Processing an expenses claim while on the move, using voice activation on a smartphone
- A bot proposes a meeting when a customer is due to renew a contract
- Multilingual communication between employees in global offices
- Accepting an invite while driving
- A manager asking a chatbot to find a performance appraisal form
- An employee commands the company AI to 'fetch me this month's revenue' and the AI serves up the document

Such consumer-friendly interfaces are only possible when tools and platforms are seamlessly integrated behind the scenes. APIs allow applications to 'talk' to each other, and they can also link a chatbot on an employee portal or intranet to information from relevant systems 'under the surface'. If this is done badly, the benefits of offering an easy user interface are lost, argues Henry Amm, managing director of Adenin Technologies, which builds consumer interfaces and digital assistants for intranets. "Companies often have multiple apps in production, with employees navigating anything between five and 15 on a daily basis," he explains. "You can create a bot so someone can ask about an HR policy and the bot links to the HR system. But where the experience can fall down is if it then offers them an 80-page PDF they can't read on their mobile phone. If it can give a direct and definitive answer, it's saving that user clicks and time." Organisations can use analytics to identify where requests are falling down and tweak the AI engines behind the interface. "You can tell the engine, 'Here are 100 things people might ask' or show it the words they might use for different requests. You need to teach the AI so people can ask things naturally," adds Amm. Adenin constructed a digital assistant for tech giant Cisco for employees to access their everyday applications in one place; the company has since won awards for digital workplace readiness and experienced a 17% increase in workplace satisfaction.

As AI becomes more sophisticated and learns more about the context of an organisation and how people work, these interfaces can begin to act more like digital colleagues than technology tools, says Johan Toll, chief technology officer for IPsoft, which has developed a digital assistant called Amelia. "Use cases are beginning to change from having a bot that can answer a question to [one that gets] something done. So, the user might say 'I need three days off next week' and they are not sent a form, the time just gets booked in," he explains. "If someone has lost their credit card, they can move to a hybrid model if they want to speak to a human, who then passes the resolution back to the assistant. This is what creates great experiences for employees." Open APIs make most assistants 'agnostic' to the systems at the back end they're connecting, which means organisations can customise and personalise the employee experience. Toll adds: "In the past, any kind of human-to-service case meant the user needed to adapt to the system – so if you move employer, you have to learn a new programme. By putting in a virtual assistant, this adapts to the employee rather than the other way around. This is a big change and also drives engagement."

Just as we might put a human colleague on probation, time and effort are required to educate and test these interfaces. This can mean there is a frustrating period where employees do not receive the precise answers they need, and when HR or other teams need to review and adapt the data feeding its responses. It can help to communicate with employees that the system is still learning, and ensure support teams are available to answer questions in the interim. Organisational circumstances are also very fluid: consider asking a chatbot about furlough in 2019 before the pandemic began. Questions and responses will need to adapt as legal or business developments lead to changes in policies or allowances.

"As humans, we don't do things the same way every time, but an assistant will carry out a task in a linear way each time. If you get a bad response, you won't use it again, so you need to show it can deliver value," explains Toll. He advises organisations to start with simple, real-life scenarios and build up from there. While we might assess human employees quarterly or even annually at an appraisal, AI interfaces are learning from feedback every day and adjusting their output accordingly. Janes adds: "The best success comes from having a period of experimentation – a 'fail quickly' mentality. This is why lots of organisations focus on transactional processes where it's easy to define the knowledge base, and it's still early days in terms of impact on employee experience."

The next step is empowering employees and managers to create their own assistants, Toll predicts. "A HR person is the best [person] to know how their processes work, so how do we train them to train an assistant and create their own automations? This is how we democratise AI." IPsoft has developed a 'digital employee builder' that enables subject matter specialists to piece together blocks of AI that suit what they're trying to achieve. As this market evolves, it will also be incumbent on HR to work with other parts of the business to build frameworks for how managers interact with their virtual colleagues. They will need to build workflows to show how matters escalate when the digital tool reaches the limits of its support, and manage how data is stored and protected. We'll see more specialist roles develop – such as 'chief robotics officer' – to manage such tools.

As organisations strive to create better experiences for their workers, it's important not to assume that what's 'good' for consumers will translate to the workplace. The reasons Alexa makes life easier at home may not equate to increased engagement at work, even if bots and assistants help us to be more productive. As Deloitte's report concludes, consumer interfaces need to be part of a broader push for engagement that goes way beyond technology. "Employees are different from customers: they have an enduring, personal relationship with their employers, unlike customers who can stop buying an organisation's products at any time. The employee experience is social: it is built around culture and relationships with others, moving well beyond a focus on an individual employee's needs. And most relevant to the issue at hand, employees want more than an easy set of transactions; they want a career, purpose, and meaning from their work."

Five key takeaways

- Think about your organisation's 'use cases' for a virtual assistant, chatbot or other consumer interface. What problem do you want it to solve?
- Start small and break down processes to their simplest components; this increases the system's chances of success
- HR is a natural fit for consumer interfaces – automating common employee queries is a good way to start an evidence base for more investment
- Invest time to build knowledge in your AI tools and tweak if necessary

Talk to IT and marketing departments about consumer-facing tools – could something similar work for employees?

Liberty Mutual's chatbot gets work done

Insurance group Liberty Mutual developed its own chatbot through its innovation startup, Workgrid Software, in 2018. The chatbot can handle enquiries across functions including HR, IT and facilities, and employees can access it through their phone or any browser. It can answer questions and automate common tasks, but can also personalise responses based on job function, primary office location and more. It comes prebuilt with a collection of 'small talk' interactions such as greetings and jokes, but also boasts a 'no code' interface so departments can train and deploy it in a way that suits them. The chatbot sits within a wider digital experience platform that brings together multiple applications so employees can find information, and manage their to-do lists and sign offs without having to switch systems. Employees now need 70% fewer clicks to obtain key information, and 80% of transactions are approved in Workgrid as opposed to the source systems behind it.

PART FOUR

L&D AND CAREER PROGRESSION

Managing performance

"There are three things you need to know about an employee's performance," says Tom Marsden, CEO of team coaching software company Saberr. "Are they sinking, swimming, or walking on water? If they're sinking, you need to have an honest conversation with them; if they're swimming, they might need to improve their stroke; and if they're walking on water, how do you recognise that?" For too long, Marsden argues, performance management has been overcomplicated, and focused on one formal, annual review that dwells on past achievements rather than future potential.

A review of how organisations monitor employees' performance by the CIPD[57] in 2016 emerged just as a growing number of employers announced plans to ditch annual appraisals. Netflix reportedly dropped all formal reviews, while Accenture and Microsoft stopped using a process known as 'forced ranking', which scores employees along a curve of high performance, average performance, and underperformance. The CIPD's assessment of performance reviews was that "they are seen to be overly time-consuming and energy-sapping, disappointing and ultimately demotivating for employees, divisive and not conducive to cooperation and effective team working and, most damningly, not effective drivers of performance". Research by consulting firm Deloitte echoes this, with a study of human capital trends concluding[58]:

57 Could do better? What works in performance management, CIPD, December 2016, https://bit.ly/3qn6zA5. Accessed 23/03/2022

58 The Performance Management Puzzle, Deloitte Human Capital Trends, 2013, https://bit.ly/3qpF48V. Accessed 23/03/2022

"Organisations continue to use traditional performance-management processes because they provide a consistent way to evaluate employees and apportion rewards. But when it comes to motivating and engaging people, these conventional processes seem increasingly obsolete."

So how can managers get an idea of how employees are doing and how their performance might change (for better or worse) in the future? It is possible to find a middle ground between an annual benchmark of performance. According to a survey of 48,000 employees, managers and CEOs by training company Leadership IQ,[59] only 13% of employees and managers think their organisation's performance appraisal system is useful. And while a vocal handful of employers have ditched the annual review in favour of a more fluid approach, it's fair to say that most organisations will still rely on running an annual review of some sort. That said, one of the recommendations of the CIPD's research is that more regular feedback mechanisms help employees feel more motivated because they can see progress towards a goal, and how their contribution fits into the wider goals of the business.

"Managers should be giving feedback continuously, behaving more like a coach than an assessor," says Roly Walter, founder of Appraisd, a performance-management technology company. He believes there can be a number of biases inherent in using ratings or rigid appraisal processes, such as employees feeling underappreciated because their manager tends to give lower ratings than others, or managers' subjective opinions of the individual concerned. "You can try to remove bias by adding in objective data such as sales targets, but anything that's associated with personality can become tricky," he adds. Indeed, the CIPD found that perceived fairness was one of the key success factors in any performance management approach – that if employees felt they had been unfairly treated, the mechanism did not matter.

59 Performance Appraisal: New data reveals why employees and managers dislike them, LeadershipIQ.com, https://bit.ly/3zXq6Kx. Accessed 23/03/2022

Technology can support managers to deliver fairer and more regular feedback on performance. It can nudge managers to book time with one of their team for a progress update, it can collect data from performance meetings, and it can enable managers to share someone's success more widely over a company intranet or messaging platform. "If you want to make your organisation more feedback-friendly, you need to improve the lines of communication across the business so it's less siloed and more agile," says Lee-Emery from Head Light. Technology tools can also gather a broader range of perspectives on someone's performance, such as soliciting feedback from their peers, or asking employees to rate their own perceptions of success in their role. "In the past there has often been undue emphasis on what the manager thinks because they know the person well and they're there to support and coach them," adds Lee-Emery. "Asking people to build a bigger picture of themselves – for example, asking peers whether their communication skills are up to scratch – can help break down those barriers."

One of the positive developments of the pandemic is that managers are checking in more regularly with employees who work remotely. Many tools include functionality that can prompt managers to have these conversations, or even suggest areas for discussion based on performance data in the system. "With the shift to remote teams and hybrid models, there's been a big trend towards organisations trying to solve problems in a more collaborative way, and giving more power to local teams to make decisions on the ground," adds Marsden. "This means the leader acts more as a coach to the team." Historically, much appraisal technology has focused on individuals' assessment and performance, but this is gradually shifting to recognising the work that is done within teams, he says. Nudges within Saberr's tool CoachBot prompt managers to set up individual and team feedback sessions, and include suggestions for learning content or discussions linked to the company's values or future goals. "The pandemic means we don't have those informal feedback conversations anymore, so we need some more thought and structure to our conversations," he explains. "Technology can help you create rituals and routines that then become habitual."

Dawn Baron, director of marketing at recognition software company PeopleFluent, believes that using tools to record and stimulate feedback can support managers to recall where employees need support and celebrate successes. "Even the best managers struggle with recency bias, as it's difficult for managers to accurately remember what each of their employees has worked on over an extended period of time," she says. "Having a space [where this is recorded] offers a single source of truth for performance management. Managers can build a holistic picture of employees' performance by pulling in relevant data from employee feedback, recognition, goals, and agenda items from past 1:1s. These insights can help foster a discussion around key accomplishments, learnings, and expectations moving forward."

Linking up data from performance conversations to learning systems and other HR tools has the potential to create a virtuous circle. Employees could be recording career goals or achievements in an organisation's talent network platform, performance discussions might identify areas where skills need to be built to reach those goals, and learning content can be created and curated based on what the workforce needs. Perry Timms, an organisational development consultant and founder of PTHR, says it's crucial that performance management does not sit in isolation, particularly as the size of the remote workforce is growing. "They're not always getting clarity around what's expected. Linking performance management to learning means people can 'size' their work better, so if you have some people who are busy and others that are underemployed, can learning help you redistribute that load?" Greater control over how they shape their role now and in the future can help breed higher engagement and a better work–life balance, he adds.

When selecting or implementing performance management systems, there are several considerations as to what elements it might include, such as:

- How frequent will performance conversations be?
- Will you have an annual 'benchmark' of performance? What other elements could be tracked on a more regular basis, such as career progress and management skills?
- Will you include input from peers, self-assessment or 360-degree feedback? How will you solicit and record this?
- How will you set goals: will you base these on time (eg quarterly), a project cycle, customer feedback, or team performance?
- Will you link performance measurements to pay?

The question of linking remuneration to performance is a thorny one. Reports emerged in early 2021 that the *Daily Telegraph* would link some elements of journalists' pay with the popularity of articles, provoking accusations that this would prompt the wrong behaviours and prove unfair to those with less popular coverage areas. Lee-Emery argues that it could be detrimental to someone's motivation if any dip in performance is reflected in how their salary is set, while Marsden advises keeping conversations about pay separate or allocating pay on how a team has performed, rather than drilling down to individuals' work. In Acas's survey, only 8% used appraisals to determine base pay, and 1% for bonuses.

Looking forward, there are a few factors that influence the evolution of how we manage performance, and the technology we use to support this. Increased automation of certain roles means employees' outputs and targets will change, while many organisations may choose to use more flexible and contingent sources of labour in an uncertain post-pandemic business climate. Providing constructive and regular feedback amid such fluidity will become highly complex – and systems will need to reflect this.

In the meantime, however, the pandemic has necessitated a shift to more deliberate feedback conversations because most employees are being managed remotely. Baron from PeopleFluent concludes: "Without seeing the nonverbal communication, it can be hard to know how someone is feeling in a daily meeting, let alone when receiving feedback. Being remote shouldn't change your approach or strategy to performance management, but it does require us all to be much more intentional with our communications." Prompting and informing these conversations is the key role for performance management and feedback tools, and one that is not likely to disappear.

Five key takeaways

- More often means better engagement: regular conversations about performance help employees see how to meet goals
- You don't have to ditch the annual appraisal: it can be a useful benchmark for employees to refer to
- Rankings can disengage employees who feel they don't fit an ascribed performance curve or who may have other valuable skills to offer
- Simplicity is key: making performance management too complex means managers will avoid it, and employees can't as easily see how their role fits into organisational goals
- Adapt to a fluid workforce: think about how you offer feedback to 'non-traditional' employees such as freelance workers, or adapt performance management to hybrid and remote working

Jo Faragher

The importance of good talent management

'Mary' has all the qualifications and experience for an upcoming promotion, and is ready to take the next step in her career. Yet when she is offered a promotion she turns it down, and joins a competitor a few months later. What went wrong? Her former employer expected her to relocate to another country, which was impossible for her due to family commitments. Her friends knew, but HR didn't – so the business missed out.

"We can store information on assessment, previous work history and feedback data on our talent systems but if all that information goes into the system to die, there's no reciprocal value," says Roger Philby, CEO and founder of workplace consultancy Chemistry Group. Talent management and planning, he argues, should be a fluid exercise where all parties contribute and actions happen because of the insights the technology generates. "If I update my personal development plan after some training and I'm pinged an opportunity a few days later, it shows it's worth it," he adds. With the shape of work rapidly changing due to the twin forces of a global pandemic and increased automation, having access to a 'living' talent map that reflects the skills and competencies available now – and in the future – will be a crucial tool that determines how organisations survive and thrive.

"On talent, we see two trends," explains Dorothee El Khoury, HR practice leader at analyst company The Hackett Group. "Even before Covid, there was a need for the talent cycle to be integrated and for a better understanding of strategic future workforce needs – not just headcount, but also skills. This needs to feed all the talent management processes: from sourcing, recruitment and training, to performance and planning. Secondly, HR processes are too often happening in silos, with people recruiting for open positions but not thinking about mobility, while training is following a different logic altogether. It's not joined up so there's a lot of wasted effort, and it means HR is not aligned with what the business needs." The answer, she says, is better integration between learning, central HR, performance management, and recruitment systems so organisations can better match open opportunities to people who have the skills, potential and desire to fill them.

Many organisations will be familiar with talent pools already – essentially a database of potential job applicants that have already shown interest in working there or who may have applied before and been unsuccessful. But this approach tends to focus on acquiring talent and bringing it in from an external source; it may save money on recruitment advertising or using agencies, but it's still bringing in someone from outside. "We see a lot of technology innovation at the talent acquisition funnel," says Simon Lyle, managing director of Randstad RiseSmart, an outplacement provider, "but then it tends to spring a leak. In the future there will be more use of flexible talent as organisations need to deliver smaller projects very quickly, and this could be a mix of contingent workers or people who already work in the organisation." Many larger companies, particularly in professional services and consultancy, have responded to this by building their own talent 'marketplaces' that they use as projects require.

Building a flexible talent pool that can adapt as the needs of the business evolve could include:

- A database of people who have shown interest in working for the company before (including sign-ups from sources such as a careers page or LinkedIn)

- Data from feedback discussions with existing employees on their career aspirations
- Results of psychometric assessments detailing behavioural strengths and weaknesses
- Integration with an LMS so content can be curated to support employees to meet their aspirations
- A record of learning completed by existing employees
- Up-to-date details of company alumni who may be interested in projects in the future
- Data on freelance and contingent workers with details of specific experience or qualifications
- Details on people identified as high-potential workers or potential successors

Whether an organisation incorporates all these components, or just some, the key is to ensure the information in it is up to date, and that requires a commitment from employees, managers, HR and leadership teams. This is where many organisations fail, according to El Khoury. "What companies tend to do is segment different areas of the workforce and have different talent pools, rather than looking at the workforce as one." Instead, organisations should take more of an 'ecosystem' approach, she advises, and the components could be any type of worker from full-time staff to agency workers, collaborations with start-up businesses, and independent workers who sometimes work directly for a company and sometimes through an app or platform. "The challenge here is to have visibility of this ecosystem – where the dependencies are, where the risks are, who has intellectual property, and how policies apply to all of those parties," she says. "This means HR, research and development, procurement and legal departments all need to work together."

Despite an uncertain labour market, many organisations face skills shortages in key areas, such as technology, as companies accelerate their digital transformation plans. According to Jo-Ann Feely, global managing director of innovation at recruitment outsourcing company AMS, there are 2.7 million open tech vacancies worldwide where demand is not being met. "The pandemic has brought the future closer to us and businesses' digital transformation journeys are playing out in the numbers. All businesses have had to pivot and they've turned to digital platforms," she explains. When it comes to managing digital talent in a tight market, this requires what she calls 'workforce dexterity'. She adds: "Rather than thinking that you need to add or replenish talent, think about how you upskill and reskill your existing workforce. Screen people for potential and they will have enhanced mobility options internally." AMS takes this approach with its graduates, who are hired for their potential and then trained in specific digital skills as and when necessary. "This means you can reshape the workforce so it's the tech talent you require. It stops you thinking 'We don't have these skills, let's recruit a contractor' – instead, you offer those opportunities to your internal population," says Feely.

Organisations will still of course need to rely on contingent workers for particular projects, and using technology to support talent management means they can have oversight over where these skills are coming from. According to the Workforce 2020 report from Oxford Economics and SAP,[60] 83% of executives plan to increase their use of 'contingent, intermittent or consultant' employees in the next three years. However, only 42% say they know how to extract meaningful insights about those workers. Lyle at Randstad RiseSmart believes businesses will need to 'open up borders' to their contingent workforce to create higher levels of engagement and loyalty. "From an engagement perspective, you can redefine engagement policies that had previously been targeted solely at your permanent worker population," he says. "Think about the different forms a worker can take; can you, for example, open up online platforms so they can engage with you?" Rather than locking

60 Workforce 2020: Building a strategic workforce for the future, Oxford Economics, 2020,
https://bit.ly/3I48NKN. Accessed 23/03/2022

this worker population out of platforms such as learning management systems, inviting them to boost their own skills is a win–win, he adds.

Becoming more 'borderless' in how you map talent can also be beneficial in terms of equity and inclusion. The push to remote work means that geographical barriers can be removed for some roles, while increased flexibility over hours and location opens up work to groups that may have felt excluded before, such as working parents or those with disabilities. At AMS, bringing in talent based on potential and helping employees reskill throughout their career has enabled the company to hit workforce targets around gender, ethnicity, and social mobility. A talent system that can make intelligent, automated suggestions for projects can also get over our human propensity to 'go with who we know', adds Philby from the Chemistry Group. "We know that often the noisiest people are the ones who get the roles. So, when an opportunity comes up, the engine will recommend someone without them having to put their hand up. We know that women tend to not apply for projects if they don't feel they tick all the boxes, and a good talent engine will remove that barrier."

But, with the McKinsey Global Institute predicting that around 30% of work done could be automated by 2030, how can we incorporate our new 'robot' colleagues into talent planning and management? El Khoury believes organisations will need to rethink the skills that employees require and this will define how they map talent. "The workforce will be a hybrid of automation and people. But while robots require little emotional management, their human colleagues need to be aware and be able to react if something goes wrong – think of how we use the autopilot function on a plane, for instance," she says. Furthermore, AI tools will not be able to perform higher-value tasks such as dealing with complex customer queries. Feely adds: "We still need people to do the engagement work. We can get a basic legal contract from an algorithm, but you need someone to do the relationship management." A robust talent management system will provide oversight of those softer skills, or link to a learning system that can curate courses where there are gaps in this area.

Recruitment company ManpowerGroup argues that organisations' priorities post-pandemic should be to 'renew, reskill and reboot', identifying where their existing workforce can learn new skills and developing soft skills so employees can cope with changes to their roles in years to come. Talent management can support HR teams to manage the ebb and flow of skills both within their business and among external workers, save money on recruitment, and retain an engaged workforce.

Five key takeaways

- Go beyond a talent database: integrate former candidate data with information on current employees, performance management and learning systems
- Think ahead: hire for potential. Functional and soft skills can be learned and developed over time
- Work across departments: collaborate with procurement teams, line managers, finance, and IT to ensure data is up to date and secure
- Use talent management tools to reduce human bias and increase diversity by automating suggestions for roles
- Incorporate untapped sources of talent such as alumni networks and contingent workers rather than relying on external applicants or existing employees

The power of alumni networks

Another strand of talent that should not be ignored is your organisation's alumni. Many larger employers have established alumni networks with thousands of former employees, often in the form of an online portal where they share important company news and advertise face-to-face events. But while they're a great brand engagement tool for former colleagues, they can also potentially be a goldmine for talent.

In February 2021, retailer Marks & Spencer (M&S) launched its first official alumni network on a digital platform. Members can keep up to date with what's happening at the company and access to live job opportunities. There are interviews with other alumni and 'nostalgia' content from the retail giant's 130-year-plus history. To increase its reach, the network is being pushed across social channels such as LinkedIn and store community Facebook pages.

Katie Bickerstaffe, chief strategy and transformation director, says the network is made up of "a huge community of former colleagues who remain our most passionate advocates and constructive critics." The network, known as M&S Family, will not only offer access to a wealth of potential temporary and permanent workers, but also offer support for community initiatives and mentoring, she adds.

Upskilling and reskilling the workforce for an uncertain future

The past couple of years have been a perfect storm for workplace skills. The World Economic Forum was predicting even before the Covid pandemic that 50% of employees would need reskilling by 2025,[61] while months of economic shocks, remote working and furlough have shaken up the labour market beyond recognition. Consulting firm McKinsey describes this as a 'double disruption'[62], where the combination of increased automation and changes to working practices means organisations need to know where skills are, reskill people where appropriate, and support employees to acquire skills where there are gaps. Its research found that 87% of executives felt there were gaps in their workforce skills or that they expected there to be within a few years.

McKinsey argues that in order to thrive in a post-pandemic world, organisations need to take six steps:

- Identify the skills that your recovery model depends on
- Build the skills critical to your new business model

61 We need a reskilling revolution. Here's how to make it happen, Børge Brende, World Economic Forum, April 2019, https://bit.ly/33iWwDt. Accessed 23/03/2022

62 To emerge stronger from the COVID-19 crisis, companies should start reskilling their workforces now, McKinsey &Co, May 2020, https://mck.co/3fpud8Y. Accessed 23/03/2022

- Create tailored learning journeys to close critical skills gaps
- Become more 'agile' and take on a small company mentality
- Test new approaches and iterate
- Protect your learning budget

In short, this means taking a step back to consider how the goals of the business might have changed, what new behaviours and skills this might require, and adapting approaches to delivering them. Timms of PTHR advocates 'disassembling' jobs and putting them back together in a way that reflects changing requirements. Technical skills are a good example here, as even non-technical roles require increasing levels of digital competence and understanding. He says: "You have to look over the fence and see what the next set of skills might be. You might need banking experts who can code or ways to take people out of their specialism." Supporting people to broaden and deepen skills that will allow them to 'cross-cut' across roles is also crucial as companies consider how talent is deployed. So good project managers can move across departments, for example.

This approach will become essential as organisations look to rebuild as the economy recovers. And in some cases, rapid reskilling of employees will help to avoid redundancy in the longer term. "Upskilling and reskilling people into growing industries is crucial to combat ongoing workforce displacement," says Chris Gray, director at recruitment company ManpowerGroup UK. Take retail banks: as branches closed during the pandemic, employees needed to build their empathy skills as virtual consultations with customers surged, or to deal with a rise in applications for specific products such as more flexible mortgage options or insurance. Lyle from Randstad Risesmart argues that companies will need to develop 'always-on' skills development offerings that will help employees to adapt their knowledge as jobs change or new roles emerge. "It all starts with awareness – employees holding a mirror up to themselves to find the skills gaps they need to address," he says. "Career development is a huge driver of engagement." Randstad RiseSmart's global skilling survey recently found that 70% of HR teams asked or required employees to upskill or reskill to meet changing business needs. Another survey, by compliance training company DeltaNet, found a 70% increase in searches related to career switching between 2021 and 2017 as workers reconsidered their career priorities during the pandemic. "It's going to become more important than ever that new and long-term employees are offered more than 'just a job', but a real future with ongoing opportunities for upskilling," says managing director Darren Hockley.

The way organisations build skills, as McKinsey suggests, will also need to change. Tech companies and start-ups are used to working according to 'agile' practices, which involve short sprints rather than long projects, and where adapting things along the way is all part of the process. Timms adds: "We need to stop looking at work in a big, unpalatable way and instead stage things in increments, breaking them down." Learning will also become increasingly bitesize and 'just in time', with upskilling employees to become more agile in their approach becoming an increasingly central part of learning strategy. One of the challenges, however, is that employees can be swamped by choice if learning content is not curated to their individual needs. "A learning advisor or introducing mentors that can challenge employees' choices is the bit that a LMS can't do," says Lyle. "They can then carve out time in their day to fill their learning gaps and nudge forward."

Giving employees more autonomy over skills development can also boost engagement. Digits' learning platform, for instance, offers a skills advisor tool where employees can rate themselves against certain competencies. For example, if they want to build their teamwork skills, they can complete a questionnaire that will identify courses that will help them achieve that benchmark. "They can show you where you are compared to others in your department, how different people in your organisation rate themselves, and the results can be used for performance conversations," explains Toby Gilchrist, head of implementation services – LMS, at Digits. Skills analysis can feed into the company's changing needs, show where the gaps are and aid succession planning, he adds. "You can see where people can move if they change their skills, and where people might need to be upskilled for management and leadership." Linking skills with career aspirations and plotting a pathway for employees can not only boost their confidence in learning but help the organisation to build a snapshot of talent 'now' and how it could look in the future.

With how and where we work likely changed forever by the pandemic's impact, management skills will need to be high on organisations' agendas. HR technology consultant Denis Barnard believes the erosion of management skills had been happening for some time pre-pandemic as people were promoted for their technical capabilities rather than their ability to manage others. "Management skills were missing anyway but now they're managing a hybrid workforce, the skills gap will go through the floor," he says. "For too many organisations this has been on their radar, but they've been waiting until people could come back to the office. Managers need to learn to be empathetic, to work in a collaborative way, and to guide people to find the best way to get a job done. It's not just about knowing the job better than anyone else."

Another side effect of so much work being pushed into the virtual space is that the use of productivity tools such as Microsoft Office 365 and Slack has soared. This has forced many organisations to realise that employees' digital skills were lacking, according to Adam Lacey, operations director for Simon Sez IT, an IT training company. It can mean that employees spend much of their day working out how to do simple tasks in tools such as Excel rather than focusing on deeper knowledge work, which drives down engagement and increases people's stress. "The usefulness and need for these programmes has been heightened, and people are spending more time using them," he says. "Employees need to learn how to become productive when using them rather than burning out, and the first step is learning how to use them properly." While some of these everyday tasks will be automated in years to come, investing in basic digital skills improvement pays off because employees can focus on where they add value.

Heather McGowan, a consultant on the future of work and co-author of *The Adaptation Advantage*, says the way we acquire and transfer skills will evolve as automation increases and roles disappear or are created. "The problem is we've been using models from the past. It used to be that if the market needed a certain skill you'd have around a decade to build it, acquire a degree or put together a professional qualification," she says. "We don't have a decade anymore, and the shelf life of skills – particularly in technology – is now around two or three years rather than five." Moving forward, the job for organisations will be to prepare employees for this constant change, both in upskilling them and in supporting them to gain the resilience this adaptability will require. She concludes: "We need to get better at human capital; how we reset people's expectations, how we help people to adapt, and that we need to reskill throughout our careers."

Five key takeaways

- Put learning in employees' hands – allow them to do their own skills analysis and link to career aspirations
- Build management and soft skills as work becomes hybrid or remote in the longer term
- Invest in improving everyday technical skills so employees can spend more time on high-value activities
- 'Disassemble' jobs so you can see what the components are and where skills need to be improved, or how skills might change as the role evolves

- Adopt a lifelong learning and reskilling strategy that helps employees to feel comfortable gaining new skills

Case study: Jardine Motors Group

Jardine Motors Group is an automotive franchise group with 68 locations across the UK. Historically it had used face-to-face learning across its dealerships, but this was both expensive and time-consuming. Employees also had to restrict their learning to certain times rather than being able to go at their own pace.

In 2018, Jardine turned to Digits to provide eLearning for compliance training on the General Data Protection Regulation (GDPR). One of the attractions of Digits LMS was how it presented learning, with regular email campaigns letting employees know what's new and homepage content refreshed fortnightly.

Digits LMS really came into its own during the pandemic. The learning team had been making plans to push more content onto digital platforms for some time, but this was accelerated during 2020. In the first lockdown, the company needed to place around 95% of employees on furlough. Employees on furlough could continue to access the LMS to build skills or look at wellbeing content, while around 100 employees who were still working, albeit remotely, were able to use their downtime to build their skills.

OLI (online learning innovation), as the LMS is known to Jardine employees, has been instrumental in building employees' confidence as they return to work. When showrooms reopened in summer 2020, employees could access videos about how the workplace would look with Covid-safe measures in place, making them less anxious about the return. A gamified 'digital ninja' programme has helped upskill employees in Microsoft Office 365 via a series of 'missions' for which they gain badges.

Jardine's learning team has been transformed by digital learning, increasing the number of eLearning developers and upskilling many of those who had been used to face-to-face training. In 2020 it won an award from *HR* magazine for its L&D strategy, earning top scores from judges for the way it had linked OLI to wider business goals and its commitment to measurable success.

Learning content formats and sources: what works best?

Most aspects of HR technology have undergone an evolution in the past decade, but perhaps none so dramatic as learning and development. While learning teams used to focus on designing and administering offsite training sessions with expensive instructors and a small proportion of eLearning for compliance, they now face a dizzying array of formats and tools. From tiny 'microlearning' nuggets to sophisticated virtual reality games, it can be difficult to know where to direct your budget.

The pandemic highlighted the importance of being able to access learning virtually, with the UK government offering free skills courses so those on furlough could build their skills while at home.[63] Many businesses rushed to boost their eLearning catalogues and convert in-person training into Zoom webinars or video content. Nigel Paine, a learning consultant and former head of L&D at the BBC, explains: "There was a dramatic acceleration of accessing content 'anytime, anywhere' and a shift in how organisations manage learning, with a big movement to the cloud."

According to the CIPD's 2021 Learning and Skills at Work Survey,[64] organisations' use of digital learning solutions increased during the pandemic, with 36% reporting an increase in investment in learning technologies. Just over a quarter

63 New free online learning platform to boost workplace skills, Gov.uk, April 2020, https://bit.ly/3zVDz5M. Accessed 23/03/2022

64 CIPD Learning and Skills at Work Survey, 2021, CIPD, https://bit.ly/3qrQCIW. Accessed 23/03/2022

(28%) used LMS platforms to support their content delivery, it found. These systems typically provide a mix of eLearning content and data on how employees use it, but are becoming increasingly sophisticated with features such as:

- Learner profiles that can be linked to talent management tools and performance reviews
- Reminders and nudges for essential training (eg compliance)
- Learning progress dashboards
- Recommended learning based on role, previous courses, and aspirations
- Gamified learning or 'scoreboards' for completing courses
- 'Microlearning' such as short quizzes
- Social media integration so employees can share their achievements

LMS platforms are also becoming ever more open and interoperable. There is a move towards more versatile programming standards that can support content on multiple devices and can track learner activity. This makes integration with other employee-focused systems easier and better for talent planning, and employees can access learning whenever or wherever is convenient.

At the same time, organisations recognise that learners increasingly want to drive their own learning. "Businesses are saying 'We would like you to do this, but also experiment, explore, build your own skills, and decide your own needs'," Paine adds. "The responsibility shifts from someone telling you what course you need to do to it becoming self-generated." Digits LMS, for example, helps employees drive their own learning through a tool called skills advisor. Colin Bull, director of product design at Ciphr Group, says: "You can rate yourself against competencies such as teamwork, complete a questionnaire and benchmark yourself against others. Once you've done it, you're offered courses on things where you rated beneath that benchmark – or you can use a report for performance conversations with your manager." This sort of capability will become increasingly important as the labour market shifts and employees need to keep adapting their skills to keep up, says Paine. "Circumstances are changing and people may need to pivot quickly, meaning you can't have endless intermediaries signing things off."

Learning formats

Remember VHS or Betamax? Content in learning management systems works in a similar way, and can be based on a number of technical standards. These determine how different devices will 'read' the content and deliver it to its desired audience, and are an important consideration when buying or developing learning content. Here are the key formats:

SCORM
SCORM stands for Shareable Content Object Reference Model and is the de facto standard for eLearning interoperability. SCORM tells programmers how to write their code so that it can interact with other eLearning software. It also governs how an LMS can communicate with online learning content. So if an LMS is SCORM-compliant, it can 'play' any SCORM-compliant content. It tracks courses, keeps a record of employees' progress, assessment scores, and other metrics such as time spent on screen.

AICC compliance
AICC stands for Aviation Industry Computer-Based Training Committee and is another technical standard that defines how courses interact with LMS. It is similar to SCORM with minor technical differences, such as allowing courses to communicate information in the HTTP format. Because it is an older format, some LMS platforms no longer support it.

Tin Can and xAPI

SCORM is built around the concept of standardising communications between a course and an LMS. Emerging standards known as Tin Can and xAPI cast the net further, allowing developers to send a wide range of data from other platforms, such as mobile phone apps, and share with other software such as HR systems.

Learning experience platforms

How employees learn in the workplace cannot escape the influence of how they consume media and content at home – particularly as work and home lives have become more blurred during the pandemic. Reflecting this, learning platforms increasingly offer a more 'Netflix'-style experience where they offer recommendations based on employees' previous consumption, or enable learners to consume content in bitesize chunks as they might do with social media.

This is encapsulated in the growth of so-called 'learning experience' platforms, most recently LinkedIn Learning. While most would think of LinkedIn as a job search and networking tool, in early 2021 it launched its own learning experience platform, where companies can integrate and curate their own learning content, access LinkedIn Learning's catalogue of courses, and see analytics on skills development. HR technology analyst Josh Bersin has called this a "bold and aggressive" move because it ties together recruitment and attraction with real-time skills data that can help organisations manage people better. Learning is also one of the core elements of Microsoft's new employee experience platform, Viva, reflecting the trend towards integrating learning content with other aspects of the employee lifecycle such as talent management and engagement.

Choosing content

But faced with such a proliferation of learning formats and sources, where should organisations begin? The most important thing for teams to consider is what they are trying to achieve, according to learning consultant Sarah Ratcliff. "What is it you want people to do as a result of the learning? How much time do they have to learn? What is the context and how will they use that learning in their role?" she asks. "The answer to these questions will guide you as to whether you need a video, an 'experience' such as a webinar or face-to-face meeting, or some pre-reading followed by a discussion." An example of making the learning fit the challenge is when Ratcliff worked with retailer Ann Summers during the pandemic. She helped the company to transform its online learning so face-to-face party reps could host Facebook Live parties. Pivoting the learning content and format resulted in party planning sales going up 300%.

"It's not always about the tool but how you use it, and having the time and resources to get the most out of it. You don't always need to reinvent the wheel," she adds. Curation is the key to success here: many companies produce learning content templates, for example, and off-the-shelf eLearning courses can be invaluable if you need to get people up to speed on a particular skill in a short time. Low-cost and free courses can be accessed through Massive Open Online Course (MOOC) platforms such as Coursera and FutureLearn and become part of what you offer, and informal input from subject matter experts in the business is another low-cost way to share knowledge. "Before remote working, if you didn't know how to do something you would ask a colleague – now we go to Google," Ratcliff says. "Think about what people need and how to get it to them. It's not all about eLearning courses and videos – simple processes can be written down and stored somewhere."

Paine agrees there is a shift towards 'more informal, less structured expertise' in how organisations deliver and manage learning. He adds: "Point people towards a TED talk or article, [and] let them grab what they need in an unstructured way. There's greater acceptance of multiple modes and informal conversations, as well as ready-made solutions that you can just plug in." Ultimately, however, it comes down to how the learning will be applied: an in-depth leadership course might require a combination of face-to-face coaching, micromodules on specific skills and 360-degree reviews from colleagues. In contrast, introducing a new holiday booking system to employees might only require a short video and accompanying infographic. But the real learning comes through doing, adds Ratcliff. "If I need to learn something technical, I can watch a video but I need to practise doing it – I need to know which button to press," she says. "Give people theory and examples but remember the value of social learning, too – the conversations they have that help to embed that knowledge."

Looking to the future

Isheet Bansal, head of marketing at eLearning content company Tesseract, says the most important thing is that content "remains flexible and always ready to evolve". He adds: "We map the content to the learning outcomes and then divide the content into 'must-know' versus 'nice-to-know' content. The must-know content is essential to the learning and remains on the screen. The nice-to-know content is provided in resources or other links. Impactful content enables learners to tackle new problems and seamlessly adopt new contexts." Bansal advocates that the culture, values and ethos of the organisation are also reflected in the content, and that they prioritise innovation "over and above factual content around processes, procedures and regulations."

Flexibility will become even more important as teams navigate hybrid working arrangements, argues Matthew Pierce, learning and video ambassador at learning content company

TechSmith. "Organisations want to empower learners to take control, and using formats such as video means you can create a level playing field – so if someone can't attend a session in person, they're not missing out. Or they can return to a video to get essential information if need be." Investing in formats that can be integrated across multiple platforms and share data with central HR systems enables organisations to get a 'helicopter view' of how learning is consumed and distributed, even while working remotely.

In the future, AI and automation will play a more important role in how companies curate and deliver learning, but it will still be down to the best blend that works for your organisation. Bansal sees organisations finding a sweet spot where there is a mix of in-person, virtual and eLearning delivery. Gamified learning, as well as virtual and augmented reality training, will create deeper engagement, but in addition to other formats that already add value. User ratings can be gathered and fed into algorithms so content can be targeted even more effectively. Delivery may get quicker, personalisation better and the content more interactive, but only human teams can ensure it is curated in the most impactful way.

Five key takeaways

- Investing in an LMS not only gives you a central hub through which learners can access an array of content, but also helps you monitor metrics such as completion rates – so you can demonstrate return on investment
- Keep refreshing and assessing your catalogue of content and activities; learners' needs can change rapidly
- How can you establish a learning culture and bring learning into the flow of work? People will only learn and upskill themselves if they have the time and space to do so

- eLearning isn't everything: in the hybrid world of work, organisations will need to figure out which skills and subjects are best learned through different methods such as face-to-face training, hands-on learning, and digital courses
- Make sure your learning offering is inclusive and accessible to all

Why accessibility is important

Ensuring learning formats cater for all employees is a fundamental consideration when creating, buying or curating content. Accessibility does not only cover visible or obvious disabilities such as visual or hearing impairment, but should also consider people with neurodiverse conditions such as dyslexia or autism. Not every employee is comfortable formally disclosing a disability, however, so anticipating needs rather than reacting can be a more inclusive approach.

The World Wide Web Consortium (W3C) has a list of accessibility checks[65] for online content including:

- Checking page titles accurately describe the content and differ from other pages on the site or in the course
- Checking there are text alternatives for images (screen readers will read these out for the visually impaired so they need to be accurate)
- Ensuring colour contrast works for everyone: some people need a high contrast between text and background

65 Easy Checks – A first review of web accessibility, W3C, updated April 2020, https://bit.ly/33vZggv. Accessed 23/03/2022

- Keyboard and mouse interaction – not everyone can use a mouse, so make sure people can use the 'tab' key to move to each element

- Ensure people can turn off moving content such as flashing or blinking images because this can be uncomfortable for people with attention deficit disorder or visual processing disorders

Other simple rules include providing captions when producing video content (YouTube autogenerates captions, for example) and organising content into a clear structure so navigation is consistent and easy. Seek feedback from learners on their specific needs and accommodate where possible.

According to Alex-Michelle Parr, managing director of VoiceBox Agency, multimedia accessibility is crucial if organisations are to offer a mix of learning formats. Even small changes such as being able to change the pace of subtitles or gain a full transcript can make all the difference, she says. "Accessibility is important because everyone has the right to access, utilise and enjoy the content that matters to them. Accessible content fosters social inclusion, which supports those with disabilities, those who are neurodiverse and those who are elderly."

A study by Ofcom recently found that 80% of subtitle users were not hard of hearing or deaf, and 85% of Facebook videos are watched with subtitles rather than sound. "This proves that there are a million reasons why subtitles and other adjustments may be used by everyone, not just those for whom you're making the adjustments. Inclusivity and accessibility benefits us all," adds Parr.

Press play: how games can help employees learn

One of the biggest challenges with onboarding new employees is getting them up to speed. Food chain Leon had been training up new members of staff through a combination of printed learning materials and eLearning, but had been getting mixed results. In 2020, in the midst of a pandemic, it decided to take a different approach. New employees now complete a series of minigames and interactive scenarios via their smartphones, covering menu content, health and safety, and company values. The company has been able to measure the impact, too: employees' knowledge has jumped from 50%, when they start the learning, to 92%, while 95% feel they better understand Leon values.

Krister Kristiansen, UK managing director at Attensi, the company that created the gamified learning for Leon, says that learning sticks when delivered in this format because people like to feel a sense of mastery. "People learn better when they actively engage in the content. Think about how we feel when we master a level of a video game: it releases endorphins when we reach that achievement. This creates positive reinforcement to do things the right way, and learning sticks in a way that's much deeper," he explains. More and more organisations are catching on to gamified learning as a core element of the training they offer employees: the market is set to grow by 27% a year by 2025, according to business research company MarketsandMarkets.[66]

66 Gamification market worth $30.7 billion by 2025, MarketsandMarkets, March 2020, https://prn.to/33z1cVo. Accessed 23/03/2022

Gamification as a concept has been around for some time, and the term tends to be used to describe how we apply game-style mechanics to experiences we would not normally associate with 'play'. Examples include wellbeing apps that encourage users to 'climb Everest' by getting enough steps or 'build streaks' by completing a certain number of minutes' meditation. The idea is that by incentivising someone to do the same thing again, it turns that activity into a habit. At work, this might be trying out a difficult customer service conversation or working out how to use a piece of kit. In 2019, a survey by learning management software company Talent LMS found that gamification helped 89% of employees feel more productive, and 88% felt happier at work as a result.[67]

Far from the perception that games are just for younger employees or learning noncritical skills, gamified learning is gaining traction across an ever-wider range of scenarios. At one end of the spectrum, quizzes and competitions can help embed knowledge in areas such as compliance or health and safety, while virtual and augmented reality offer the opportunity for employees to immerse themselves in experiences they need for their role. By engaging in simulations of potential interactions that might come up in their job – such as complex surgery or a difficult engineering procedure – employees can see the results of their actions without the risk that a real-life situation might entail. One example is the Ministry of Defence, which last year announced it would trial a new virtual reality training game using the same simulator engine that underpins the popular combat game Fortnite. The interactive game allows 30 soldiers to take part in a virtual battlefield combat, wearing VR headsets and holding replica guns as they move through the simulation. SimCentric, the company that developed the simulation, claims it will offer a more 'realistic, intuitive and immersive' way to experience combat without entering the battlefield.

67 The 2019 Gamification at Work survey, Talent LMS, August 2019, https://bit.ly/3rl6ZGz. Accessed 23/03/2022

It may seem counterintuitive to overload employees with more digital learning at a time when many have been spending hours on Zoom calls – technology analysts Fosway Group recently reported that almost half of organisations felt digital learning fatigue, despite the surge in demand for eLearning during the pandemic. But the incentive of scoring higher or getting something right encourages repeat attempts and high engagement with the content. Running a *Who Wants to be a Millionaire?*-style quiz to help employees know their obligations around GDPR (as clothing retailer Superdry did, for example) is more likely to engage employees than a 20-deck slideshow or lengthy video.

Bringing the virtual and the physical together means the learning can be directly applied when employees need it. Gilchrist from Digits says: "On-the-job activities capture things that will be used and assessed in the real world, and this can be good for evaluating employees' performance. Learning platforms such as Digits LMS can connect any rewards or 'badges' employees gain through their learning to HR and talent management systems so managers can map where skills may be lacking, or adapt courses based on the feedback from the data."

Having a 'safe space' in which to make mistakes and try again is one of the key benefits of learning through games. "When we relate games to learning we call them 'serious games'," says Noorie Sazen, director of learning consultancy Saffron Interactive. "They must have a purpose in terms of changing behaviour – using motivational factors and other techniques that drive that change. They can simulate environments where we have to make decisions, or come across dilemmas we might face at work. Depending on those decisions, certain consequences apply, but because it's digital you can play [the scenarios] over and over again."

But while badges, scores and league tables can add an element of fun to the learning process and support how organisations measure success, it's important not to add these for the sake of it. "It needs to be relevant, personal, and there needs to be emotional investment," adds Sazen. "Digital games can be enabling but the overall strategy has to underpin them." In building people's tolerance for failure they can acquire softer skills and learn to adapt, too. "Organisations worry about people losing motivation when they lose but they learn a lot more that way than if they get a high score. Why was I kicked out of the game? What decision did I get wrong? Denying people the chance to learn from their mistakes is counterproductive," she says.

While gamified learning gained in popularity during the pandemic as a way to train up new and existing staff remotely, it also works well as part of a blended learning strategy. Organisations can use it before an in-person training session to bring learners up to a similar level so the classroom training can be delivered more efficiently, or it can be used afterwards (or even during) to reflect on what has been shared. Kristiansen says: "We have clients that use it before, during and after classroom learning. It can ensure you have a minimum common denominator if users play before, or people can discuss their hypothetical experiences. You can use the eLearning to reinforce the high-level messages from the class and people can see how they compare to their peers – the bits they're good at, and where they struggle." Delivering short games or quizzes via smartphone apps means that employees who are time-poor can access them on-the-fly, compared with traditional learning approaches that require employees to book time away from their usual tasks to attend a course. Time spent on in-person learning can be reduced but also maximised, Kristiansen points out.

With virtual reality (VR) and other types of immersive learning, many aspects of that in-person experience can be replicated. Cost had been a barrier in the past, but the hardware required for this is becoming cheaper and more accessible. VR headsets still cost an average of £250 to £300 each, but this can work out more cost-effective than sending teams of people on an off-site training course. "It does become more cost efficient as you begin to roll it out at scale," says Sophie Thompson, co-founder of VirtualSpeech, a training platform for soft skills. "Compare it to flying people to the same location for training or a night in a hotel; you can have a few headsets employees can borrow and you get the cost back immediately." Virtual Speech helps individuals build skills in areas such as public speaking or communication by allowing them to 'experience' that important sales presentation or conference appearance. Artificial intelligence gathers data points about speed, delivery, pauses and other elements so they can receive feedback on their performance and target how they improve. "They have the freedom to make mistakes without consequences or the embarrassment of losing a big sale," she adds. "It also enables them to quantify their progress in a way we haven't been able to for soft skills before. It's really motivating for people to see their scores improve if there's an area they needed to work on."

Because games are so effective at promoting behavioural change, they are increasingly used to embed values and culture. One of Saffron Interactive's most effective projects was with tyre manufacturer Michelin, for example. The company discovered that its most famous asset – the Michelin Man – was being used to advertise off-brand and inferior products, at a potentially huge cost to its reputation. It developed a blended learning programme that included eLearning, face-to-face training and finally a TV-show-style contest at sales conferences, designed to help sales managers understand how dealers could be eroding brand value. This had financial implications for them because cutting ties with certain dealers would mean less commission, so the learning had to encourage them to do the right thing. "The quiz game [created] an emotional investment around the fact the tyres have been shown to save lives," she explains. "When they measured brand value again, it had increased by $1 billion, with roughly $100 million of this attributed to the dealers' training programme."

Just as gaming can help to embed values and culture, it can't 'fix' a culture that is not ready to learn. Introducing gamified learning needs to be part of a wider learning strategy that complements the goals and values of the organisation, rather than being bolted-on because managers think it will be entertaining or fun. Done badly or for the sake of it, gamified learning can end up disengaging employees. It can further embed current (undesirable) habits because they are turned off by the learning and dismissive of it. Sazen concludes: "If it's too generic learners can't relate, so they need to see themselves reflected in the learning, or it won't embed the kind of behavioural change your organisation wants."

Five key takeaways

- Know your strategy and use gamified learning to underpin it. What outcomes do you want to achieve?
- Think about how games can be part of a blended learning strategy, embedding classroom learning or preparing learners
- Use simple nuggets such as quizzes or short games so time-poor employees can learn on the fly
- Make gamified learning relevant. Learners need to see how to apply what they're doing in their day-to-day role
- Don't dismiss virtual reality as being too expensive. Hardware is becoming cheaper, and VR can be more cost effective than in-person training

How will AI impact learning?

Five years ago, the US Defense Advanced Research Projects Agency (DARPA) trialled the use of 'virtual' tutors to train up new recruits. Based on AI, the tutor would replicate the interaction between an expert and a novice, reducing the time it would take new navy recruits to get up to speed in technical skills. The trial was hugely effective – recruits who learned from the digital tutor were found to frequently outperform experts with 7-10 years of experience, both on written tests and when solving real-world problems. [68]

AI plays an increasing important role in our daily lives, whether it's voice assistants such as Alexa sourcing a weather report or an algorithm in Netflix suggesting a new series to watch. On a basic level, AI algorithms trawl through vast amounts of data to provide responses to questions or carry out simple tasks such as adding an item to a shopping list. Most learning and development teams will be some way off developing AI tutors to lead corporate training programmes, but that's not to say that AI does not have an emerging role in how organisations share knowledge. This is already happening in a multitude of ways, such as:

- Suggesting eLearning modules an employee might find useful (based on what they've done before or other parameters)

68 Preparing for the future of artificial intelligence, US National Science and Technology Council Committee on Technology, October 2016, https://bit.ly/3FwMuLT. Accessed 23/03/2022

- Adapting content or assessments based on how the employee performs (a quiz moves on to a different question depending on the previous answer, for example)
- Generating captions for people who are hearing impaired, or providing text to screen readers for the visually impaired
- Auto-tagging learning content, making it easier to manage and curate
- Providing answers to questions via virtual assistants (such as chatbots and Alexa-type devices)
- Offering learning at the point of need based on context (the system sees the employee has a sales meeting, so offers a short course on sales techniques)

Far from seeking to completely replace the human aspect of training, in many ways AI can augment it or at least make it more cost-effective. “In the race to deliver impact in a post-pandemic world, organisations will need to differentiate themselves from their peers,” says Sean Farrington, senior vice president for EMEA at technology skills platform Pluralsight. “Intelligent technology skill development, backed by AI, will provide an opportunity to expedite go-to-market times and deliver a competitive advantage.” This can happen in two ways, according to Jonathan Crane, chief commercial officer at IPsoft, manufacturers of AI tool Amelia. On the one hand, a digital ‘colleague’ can take over mundane tasks and simple transactions, using machine learning to improve its performance; on the other, it can enable employees to learn through doing because they are supported. “If you want to get better at something, you take the people who know what they’re doing. AI is constantly learning so can pick out material where you might not know the answer, or help you to prioritise a response. Collaboratively you’ve learned and augmented your skills, and this is the biggest benefit of AI,” he says.

Personalisation is a key element of this. Similar to the way consumer sites such as Amazon and Netflix make recommendations to us based on items or programmes we have viewed before, LMS platforms can use AI to offer recommendations of courses or content that might be useful. The delivery mechanism for this could be in the form of a chatbot or in the way the courses are presented, with a 'recommended' stream prominent on the opening page. Gilchrist says this can help employees understand where their role fits into the business, and how learning can support that role. Digits LMS, for example, gathers data on employees' skills through a questionnaire, which benchmarks them against others in their department or in similar roles across the business, which all contributes to personalising individuals' learning experiences. "The guidance is important for the self-led aspect of learning – recommending what might be relevant for them and nudging them into building their skills," he explains. Businesses can introduce parameters for recommendations based on insights from connected talent management and HR systems: so if someone is on track for a promotion or has shown interest in a particular career path, this can feed into the courses they see. AI can trigger reminders that someone needs to complete annual data protection or other compliance training, saving time and administration for the HR and learning teams.

Adaptive learning is another way in which AI can personalise content and ensure the right knowledge gets to the right employees at the right time. This could be as simple as a quiz that uses a basic algorithm to ask more advanced questions based on a certain response (or to reinforce knowledge where someone has failed to grasp it), or a sophisticated gamified learning programme that offers tailored scenarios depending on decisions made in previous stages. In both cases, the algorithm has learnt something about the user's level of knowledge and is able to offer appropriate learning content at this level. Laura Baldwin, president of O'Reilly, an online learning platform, says this can deliver knowledge at the point it is needed rather than all learners having to start from scratch. "We believe in structural literacy," she says. "If you know [programming language] Python you probably know basic functions in other programming languages. Rather than assuming everyone has to start at the beginning, a good AI platform will know this."

Any interaction with learning systems will of course produce data, and AI tools can use this data to further personalise course content and predict what might be useful in the future. In the background, algorithms produce data on learner behaviour that can help organisations see where learning has been consolidated and where particular individuals and teams need more support or are speeding ahead. It can also help learning teams see how the design of courses affects how employees interact with them, enabling them to tweak where necessary.

Pluralsight's algorithmic engine, Iris, uses data to build a fuller picture of individual learning needs but also those of the whole organisation. Farrington adds: "With every assessment and course completed, [Iris] absorbs information about the state of technology skills, collecting feedback and adapting as learners seek skills in new technologies. It uses this data to inform technology strategies and recommends what skills are needed to keep pace with change." Crucially, it can ensure learning investments are targeted more effectively. "When businesses do not use technology to underpin their employees' skill development, they often embrace a one-size-fits-all approach. This tends to result in training being delivered in classroom scenarios or through presentations by external consultants. It's costly, inefficient and immeasurable," he argues. Through intelligent recommendation, the ability to tag courses more easily, and predictive analytics from data, teams can curate bespoke learning experiences more simply – and at lower cost.

AI is already supporting accessibility to learning – think of the autogenerated captions on YouTube – and this is likely to become even more advanced in years to come. One of the pioneers in this area, Microsoft, has developed an app called Seeing AI that 'narrates' the world around someone in a range of languages so visually impaired people can 'recognise' colleagues or read text and websites. Algorithms can also distil information for employees who may have reading or cognitive difficulties, for example summarising lengthy articles or providing snapshots of articles for those that feel overwhelmed by information. We have become used to voice assistants such as Alexa and Google answering questions in our homes and these tools are beginning to be deployed in a workplace context, too.

The natural language processing and algorithms that support voice assistants to answer our questions are evolving, so arguably the natural next step is to develop 'robot' teachers that can deliver courses as we saw with the US Navy. Developments in public education are moving more quickly on this than in business. Swedish educational technology company Furhat Robotics recently piloted the use of a 'social robot' to educate school children on how to code, while Colombian start-up Van Robotics created an educational robot, ABii, that functions as a personal tutor to children with learning challenges. The robot monitors children's attention cues and adapts to their learning habits, collecting data on their progress as it does so. *Time* magazine named it one of its top inventions of 2020, and it is now used across schools in 20 US states.[69]

Such developments will seem terrifying to many, who will have concerns over losing the crucial human element of learning and the ethical implications of allowing a robot to determine educational outcomes. It will be some time before we see such technology deployed in corporate classroom learning, but some industries such as healthcare and engineering are already exploiting virtual and augmented reality as a safe way to offer learners immersive experiences of intricate tasks – and this is powered by AI. "Learning through experience, rather than watching as a bystander, brings about real behavioural change," explains Thompson from VirtualSpeech. The algorithm monitors how employees behave in certain theoretical

69 A high-tech tutor, Van Robotics ABii, Time Magazine, November 2020, https://bit.ly/3I5ycUw. Accessed 23/03/2022

scenarios and adapts the learning based on behaviour. "It measures audience perception so if you're coming across as argumentative it might suggest an in-app tool on developing better eye contact," she adds. "AI helps guide the learning and personalise it."

Looking to the future, AI will play a bigger part in scanning the horizon for potential skills gaps and augmenting what human teams do, rather than replacing them, argues learning consultant Nigel Paine. "It will always be used to help define individuals' needs and preferences, but there will also be a much bigger role," he explains. "It will provide early indicators of skills issues before they arise, package learning for individuals, and help them to make better decisions with more access to data. It will prompt us and help us to operate smarter and faster – but it won't replace us."

Five key takeaways

- AI's usefulness for L&D teams is already being demonstrated, with its importance set to grow in the coming years
- Don't regard it as a threat: think about how AI can help the human members of an L&D team make more intelligent training interventions, and cut down the time they spend on repetitive or labour-intensive tasks
- Remember, AI is only as useful as the data it is trained on. If there are flaws or biases in your datasets, algorithms will only serve to reinforce these problems
- Expect to see AI enabling more personalised learning, and adaptive learning, soon
- As AI's importance at work more widely grows, your people will need training on how to leverage it most effectively, too

Conclusion: The future of HR tech? Helping humans to be their best

Even the most evangelical technology enthusiasts would find it hard to get excited about a system that stores employees' names and addresses and makes sure they get paid every month. But in the 21st century, HR technology is so much more than a system of record.

Teams can collaborate across continents via video and virtual whiteboards; smartwatches can feed back information on whether your workers are sleeping enough; and an algorithm can recommend a five-minute training video just before your hiring manager hosts an important interview. The central HR system is still there, it's just enhanced by an armoury of tools that – thanks to the flexibility and lower cost of cloud hosting – can grow and contract depending on workforce needs. With effective integration, these systems can 'talk' to one another and the data they produce can create valuable insights – from whether someone is thinking of leaving your company to the impact of your diversity and inclusion programmes.

For anyone who's sceptical about the power of workplace technology to effect change, let's consider some of the impacts we've looked at over the course of the book:

- A sentiment analysis company ran emails from Enron through its software to test it, revealing high levels of tension that would have alerted officers at the time to the potential for a 'cultural and financial meltdown', thus avoiding one of the biggest corporate collapses in history

- A TV-show style quiz helped sales managers at Michelin to understand why selling inferior products eroded the company's reputation and could put lives at risk, enhancing brand value by around $100 million just through learning technology
- A European bank provided employees with ID badges that tracked their interactions, demonstrating through network analytics that the highest-performing branch was also the one where colleagues were most closely connected
- HR tech came to the fore in the pandemic. According to Kevin Parker, CEO of video hiring platform HireVue, US supermarket giant Walmart reduced its hiring process from 14 days to just three by using video interviews, and was conducting as many as 15,000 a day when Covid-19 restrictions began in Spring 2020

HR analyst Josh Bersin has predicted in his 'Ten New Truths about the HR Technology Market'[70] that organisations can expect layers of products and platforms to get even better at integrating jobs and tasks, but increasingly gel together employees' lives beyond the workplace. The central HR system is the foundation of this stack, with a suite of tools on top performing separate jobs such as performance management, engagement surveys and rewards management. Layered above that there's a burgeoning range of tools focused on improving employee experience as well as familiar productivity apps such as Microsoft Teams and Zoom. "Thanks to the pandemic, the number one issue in HR is engaging, supporting, and caring for people... [T]he new dimensions of remote work, hybrid

70 Ten New Truths About the HR Technology Market, Josh Bersin, September 2021, https://bit.ly/3AbIO1r. Accessed 23/03/2022

work and contingent work are on everyone's plate," Bersin says. And making sure employees don't have to log into 100 different tools just to get things done is as important as providing the tech itself.

Despite the clear benefits of having such a rich ecosystem of employee tools, there are fears that too much reliance on technology could impact jobs and wellbeing. The power of AI to automate mundane tasks, for example, has led to worries about longer-term unemployment. In 2019, the Office for National Statistics published a study showing the occupations that were at the highest risk of automation. HR administrators are considered medium risk, with a 58% chance of "some or all of the duties and tasks in this role being automated."[71] In late 2021, an all-party UK parliamentary group on the future of work called for the monitoring of workers and setting of performance targets via algorithms to be controlled by legislation, and, in some countries, this is already the case.

But others argue that the proliferation of technology in the HR space will augment, rather than reduce, the impact the function can have on the happiness and productivity of the workforce. In a volatile labour market, it can automate the transactional elements of the recruitment process such as booking interviews, but free up time for an HR manager to call the candidate, making the process feel more personal. Wellbeing data, even on an anonymous level, can prompt organisations to intervene when particular teams are suffering stress or feel overworked. Learning technology, increasingly powered by algorithms and AI, will play a crucial role in upskilling and reskilling a workforce whose jobs are changing thanks to new, post-pandemic ways of working and digital transformation. The rise of AI at work could also create jobs, as consulting firm KPMG outlined in its 2020 'Reinventing Work' report: automation of underwriting in insurance could enable people

71 Which occupations are at highest risk of being automated? Office for National Statistics, March 2019, https://bit.ly/3zUSSvv. Accessed 23/03/2022

in those roles to focus on risk analysis, for example.[72] The key lies in how HR works with the organisation to design work in such a way that this happens, according to analyst company IDC. In its 'FutureScape Worldwide Future of Work' report,[73] it predicted "a fundamental shift in the work model to one that fosters human-machine collaboration, enables new skills and worker experiences, and supports an environment unbounded by time or physical space."

Almost four decades of HR technology have shown that what it does best is to help humans to be at their most productive at work. The coming decades will not be short of challenges: shifting demographics mean managers will deal with multiple generations in one team; the so-called 'great resignation' will demand a rethink of how employees acquire new skills quickly and effectively; and teams will become more distributed as hybrid working becomes more normalised. After two years of working through a pandemic, workers witnessed a more human side to their employer, despite the crucial role technology played in keeping things running. This is the balance HR needs to sustain as it moves into a workplace future where technology plays a supporting role – but is never the star of the show.

---

72 Reinventing Work, KPMG, July 2020, https://bit.ly/3FtBhM6. Accessed 23/03/2022

73 IDC FutureScape: Worldwide Future of Work 2020 Predictions, October 2019, https://bit.ly/3ts1zfx. Accessed 23/03/2022

Lightning Source UK Ltd.
Milton Keynes UK
UKHW040611010722
405208UK00003B/95